MW01164804

VISTAS OF MANY WORLDS

A Journey Through Space and Time

Written and Illustrated by

ERIK ANDERSON

ISBN-13: 978-0-9819864-7-0

www.astrostudio.org
Ashland, Oregon 97520

Printed in Canada

CONTENTS

INTRODUCTION

How to Read the Finder Charts

Many people share a passion for travel. The thrill of immersing oneself in novel surroundings lures sightseers to far-flung regions of our diverse world. Guidebooks abound for the aspiring globetrotter. This guidebook aims at even higher aspirations. Through the itineraries presented here, we will immerse ourselves in settings far beyond our world and in epochs far beyond our times. Along the way, we'll take guidance from the stars—the common backdrop of the vistas we'll be surveying.

These journeys are imaginary ones, yet no mere flights of fancy. They probe the frontiers of human knowledge through the interplay of modern scientific disciplines. Astronomy meets geology, biology, paleontology, and archeology—just to name a few.

This guidebook is also a photo album. At each destination, we'll take in the scenery through a souvenir snapshot provided by our imaginary photographer. On most occasions, he shoots with his trusty wide-angle lens. Wide-angle portraits capture broad swaths of sky, maximizing the numbers of stars and constellations that can be framed. But professional cameramen typically keep multiple lenses in their kits and ours is no exception. On a few occasions, our photographer will opt for his super-telephoto lens. Telephoto portraits are more narrowly confined, but faraway objects are magnified and thus made easier to see.

To ensure that you always know *what* you're seeing, our photographer also happens to be a map maker of sorts. For your convenience, a finder chart will accompany nearly every photo. The charts highlight objects of interest. Whenever applicable, the brightnesses and distances of labeled objects will also be noted. The apparent brightness of each object is given in the astronomical magnitude system and distances are given in appropriate units. You'll find a quick-reference guide for both kinds of measurements on page 5.

We begin with a sample scene that exemplifies the style used throughout this book. The portrait of this ancient Roman villa on the right offers a wide field of view. Looking skyward, we note how the stars vary in brightness. The bright stars appear larger than the fainter ones, but this is a mere photographic effect. When viewed directly with the eye, all distant stars appear as pinpricks that vary in intensity. Only when viewed from close-up (as when we view the closest star to us—the Sun) can we actually perceive a star's physical size.

Let us now have a look at the chart below. It tells us that the constellation Leo is in the center of the portrait, looming over the villa. Individual stars are marked as dots, which vary in size according to brightness. The bright stars Algieba, Regulus, and Denebola are labeled, as is the planet Jupiter. Brightness and distance information (in *italics*) follow the object names. The abbreviation "m" stands for *magnitude* while "ly" and "AU" stand for *light years* and *astronomical units,* respectively.

These conventions will become more familiar as we progress through our journeys. Let us embark.

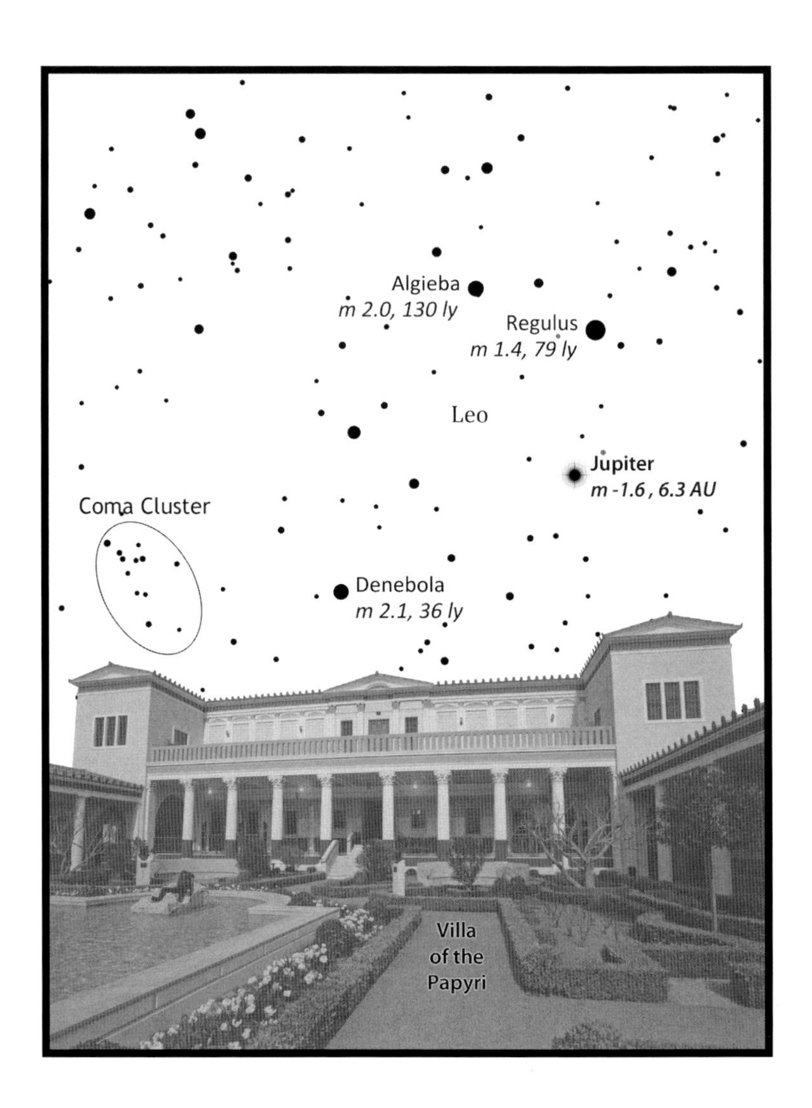

Brightnesses and Distances
A Quick Reference to Scales

The Magnitude System. A celestial object's *apparent magnitude* (abbreviated as ***m*** in this book) quantifies how bright it appears. The magnitude system was originally developed by ancient Greek astronomers, who classified stars into six ranks. Stars of the "first magnitude," the highest rank, were the brightest ones seen; stars of "sixth magnitude," the lowest rank, were the faintest. The system has been refined in modern times to give precise numeric expression. Antares, a "first magnitude" star, has a precise magnitude value of 1.06; Polaris, a "second magnitude" star, is valued at 1.97. Hence, the fainter the object, the *higher* the magnitude number.

A difference of exactly 1 on the magnitude scale represents a two and a half times difference in brightness (2.512 ×). A difference of five magnitudes exactly expresses a hundredfold difference in brightness. The modern system extends to any possible value. If, for instance, the planet Venus shines at magnitude -4, this means that it appears 100 times brighter than Antares. If Neptune shines at magnitude 8, then it is about 630 times fainter than Antares, or more than 63,000 times fainter than Venus—too dim to see with the naked eye. The following table illustrates the apparent magnitude brightness scale:

m	*Object (as seen from Earth in the present time)*
-26.74	*Typical brightness of the Sun*
-12.74	*Typical brightness of the full Moon*
- 4.89	*Maximum brightness of Venus*
- 3.82	*Minimum brightness of Venus*
-1.47	*Sirius—presently the brightest star in the Earth's sky*
0.03	*Vega*
1.06	*Antares, a first magnitude star*
1.97	*Polaris, a second magnitude star*
3.05	*Albireo A, a third magnitude star*
3.99	*Alcor, a fourth magnitude star*
5.03	*47 Ursae Majoris, a fifth magnitude star*
6.50	*Approximate limit of naked eye visibility among typical human observers*
8.02	*Minimum brightness of Neptune*
15.1	*Average brightness of Pluto*

Distance Units. The finder charts will give distances to objects in units of miles (mi), astronomical units (AU), or light years (ly). These units have the following relationship:

unit	*Description*
mi	*One mile (to convert to kilometers, multiply by 1.6)*
AU	*One astronomical unit = average distance between Earth and Sun, ~93 million miles*
ly	*One light year = distance light travels in one year, ~63,000 AU or 6 trillion miles*

FIRST ITINERARY

Ten Stars Within Twenty Light Years

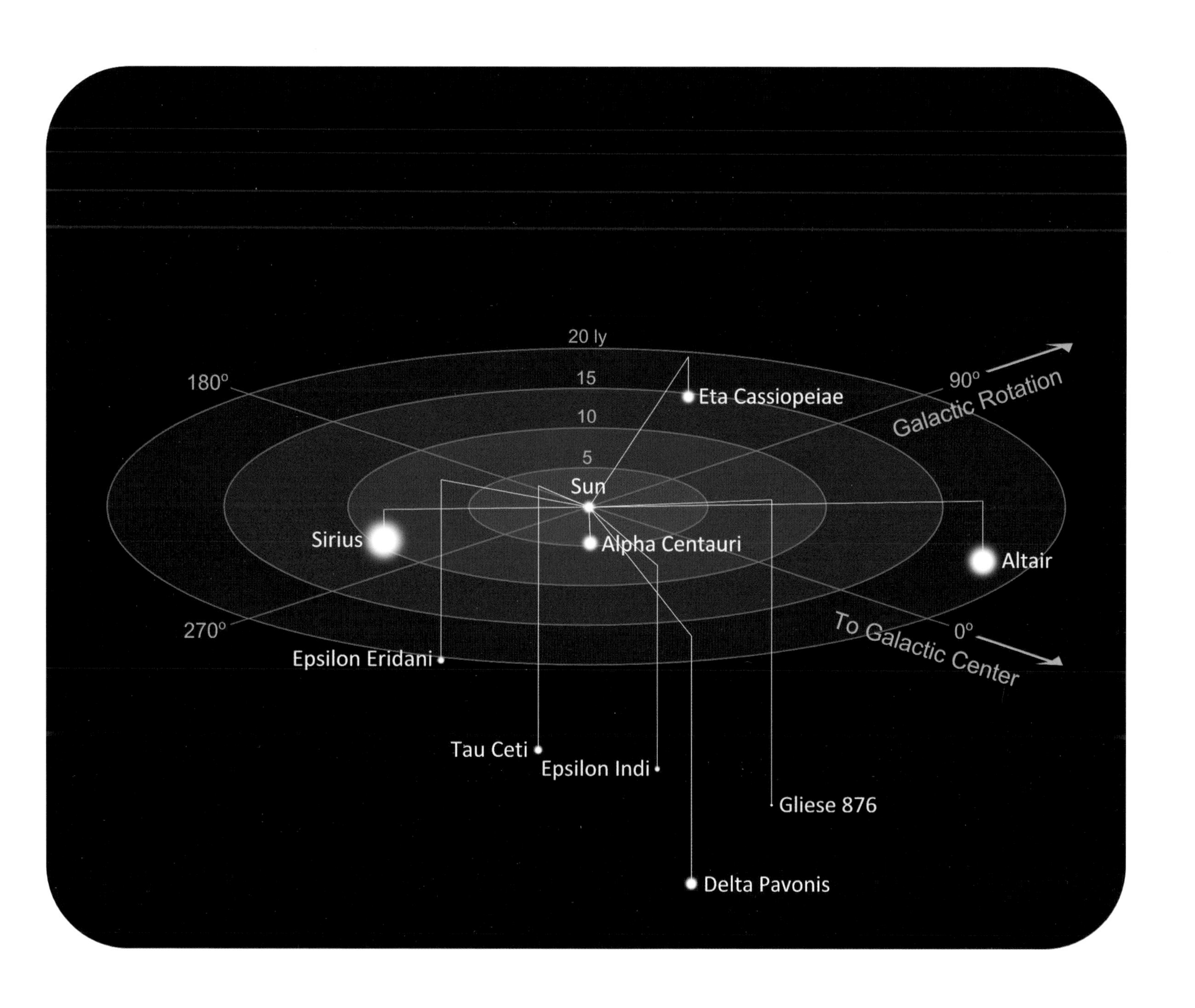

The Solar System

Homeworld of Humanity

Late one August evening in 1989, an Ariane 4 rocket departed the Kourou spaceport in French Guiana. Its payload: a scientific satellite built by the European Space Agency. The satellite was named *Hipparcos*, in honor of Hipparchus, the most renowned astronomer of the ancient Greek world.

Hipparchus had charted the positions of nearly one thousand of the brightest stars to an angular precision of one third of a degree in the 2nd century BCE. His namesake, the Hipparcos satellite, would map the positions of more than 100,000 stars to a precision better than one millionth of a degree during a three-year period ending in 1993.

As the Earth made three laps around the Sun, Hipparcos recorded how the positions of stars appeared to oscillate in sympathy. This tiny *parallax* effect indicates a star's distance—the smaller the effect, the further away the star. By applying trigonometry—mathematical tools that are another great legacy of Hipparchus—the distances of stars from the solar system can be known. Beyond about 1,000 light years, however, parallaxes are too small for Hipparcos to measure with useful accuracy.

Superimposed upon parallax motions, star positions also appear to migrate steadily in distinct directions. These so-called *proper motions* are a consequence of the various orbital motions of stars around the Milky Way Galaxy, which are different than the Sun's own Galactic orbit. Like parallaxes, the annual proper motions of stars are also too tiny to be noticed by the human eye. But given enough time, proper motions can stretch out significantly. Edmond Halley, the famed discoverer of Halley's Comet, found that the positions of Sirius, Arcturus and Aldebaran in the 17th century were undoubtedly displaced from where Hipparchus had marked them in classical times, thus shattering the age-old notion of "fixed stars."

The stars are also moving along directions that happen to line up with Earth's location—toward and away from our observed line of sight. These *radial velocities* may be surmised by Doppler shifts in spectrographs. Hipparcos was not equipped with a spectrographic instrument, but many observatories on Earth are. Consequently, most of the stars for which Hipparcos obtained useful parallaxes may also be assigned radial velocities from ground-based observations.

With accurate positions and parallaxes, one has enough information to visualize how stars are arranged in three-dimensional space. It then becomes possible to make imaginary excursions throughout the *solar neighborhood*—the region of our Galaxy containing the Sun—and view the stars as they appear from other places. With the addition of proper motions and radial velocities, we can track star motions through three-dimensional space. We can then make imaginary excursions through time. These then are the itineraries set out before us. We will explore our region of the Galaxy as space travelers and time travelers.

We begin our first itinerary by witnessing the launch of Hipparcos from our Earthly home. Stars crowd in the blackness of space. We may note the appearance of the planets Saturn and Uranus, the asteroid Vesta, and several bright stars in the constellations Scorpius and Sagittarius. Located between these two constellations is the direction toward the center of the Galaxy, roughly 24,000 light years distant. In the vastness of interstellar space, our next destination is just a tiny step away.

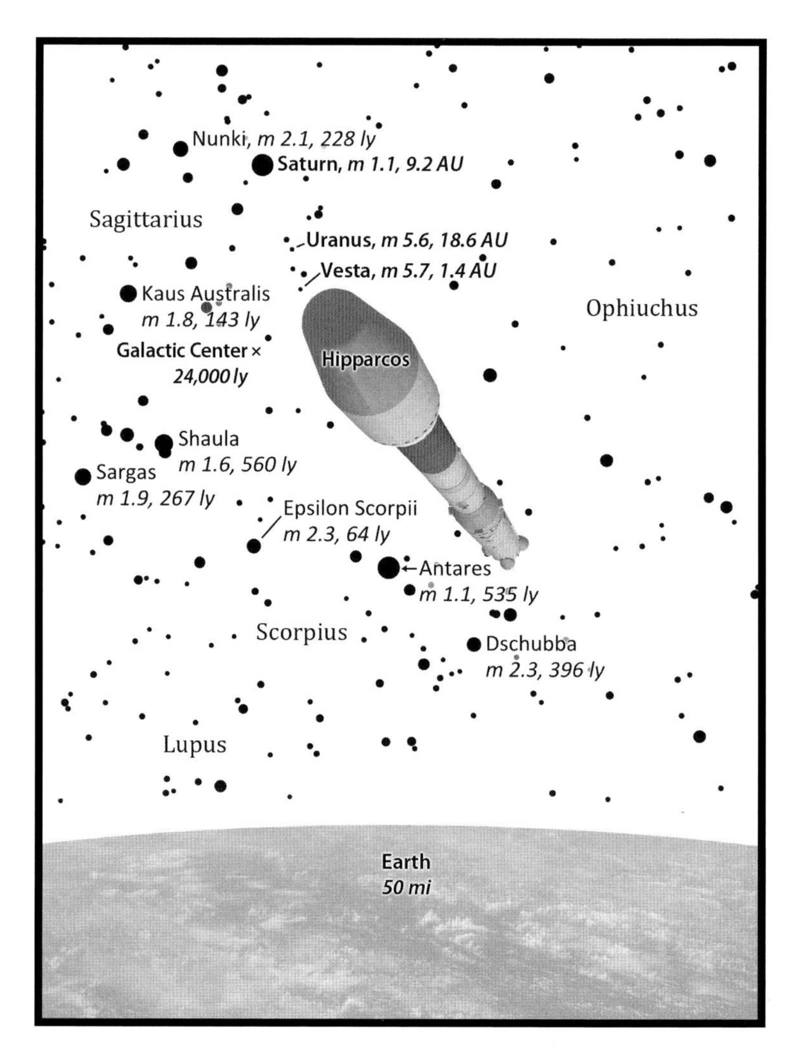

Alpha Centauri

First Milestone on the Galactic Frontier

We have just been transported to Alpha Centauri, the famous double-star system closest to the Sun. It is pictured here from our vantage point on a large moon-like asteroid with an exceedingly thin atmosphere. The main star, Alpha Centauri A, at the left edge, is slightly bigger and brighter than our Sun, while Alpha Centauri B, at the right edge, is slightly smaller and fainter.

The twin stars cast an exotic shadow play upon this desolate worldlet, but the constellations revealed in the alien sky should look familiar. Perseus stands upon the horizon at the lower right. Above him we find the famous "W"-shaped outline of Cassiopeia. Between them is the Sun, fixed in place as a bright star 4.4 light years away. At this distance, the Sun resembles Capella (lower center) in both brightness and color.

Not pictured here is Proxima Centauri, which lies below the horizon. At 15,000 AU away, this distant third companion would be a rather inconspicuous star even if it were in view. The dull red dwarf would appear at magnitude 4.8—almost as faint as the faintest stars pictured here. On random occasions, however, stellar flares may erupt from the surface of the star. These eruptions may be substantial enough to effectively double the star's overall brightness for minutes or hours at a time. Such behavior is common among red dwarfs; known examples are deemed *flare stars* by astronomers.

Alpha Centauri's proximity to the Sun has figured profoundly in the modern imagination as a primary destination for futuristic colonization. However, no planets are yet known to exist there. Alpha Centauri's double nature makes the presence of orbiting planets problematic. A planet in a system of equally massive stars could only achieve a stable orbit if it is significantly closer to one star than the other, or if it is sufficiently far away from both stars. The lonely asteroid portrayed here provides an exceptional vantage point astride Alpha Centauri AB. We may suppose that its arrival at this close-up position is made only briefly possible by a highly elliptical and potentially chaotic orbit.

The Alpha Centauri system is estimated to be 9.2 billion years old—twice as old as the present age of the Sun. It was the sheerest sort of accident that when modern civilization arose on Earth, we found Alpha Centauri to be our nearest neighbor. Its Galactic orbit follows a distinctly different path than that of the Sun. Alpha Centauri is presently migrating away from the Galactic center while the Sun is heading toward it. The two systems are navigating a celestial crossroads at a relative velocity of 32.6 kilometers per second (km/s), or 73,000 miles per hour (mph). In another 27,700 years, Alpha Centauri will pass within 3.2 light years of the Sun and then begin to recede. It is inconceivable that the Sun will encounter Alpha Centauri so closely ever again. Had our civilization emerged a few million years earlier or later than it did, we would have found the system to be much further away and invisible to the naked eye—an anonymous entry in our vastest star catalogs rather than the bright beacon of interstellar dreams.

But dream we shall, and Alpha Centauri is the most appropriate place to begin. We glance back once more at our Sun—a distant lantern hanging in an alien sky that gauges our progress. We have reached the first milestone of our journey. It is time now to voyage onward.

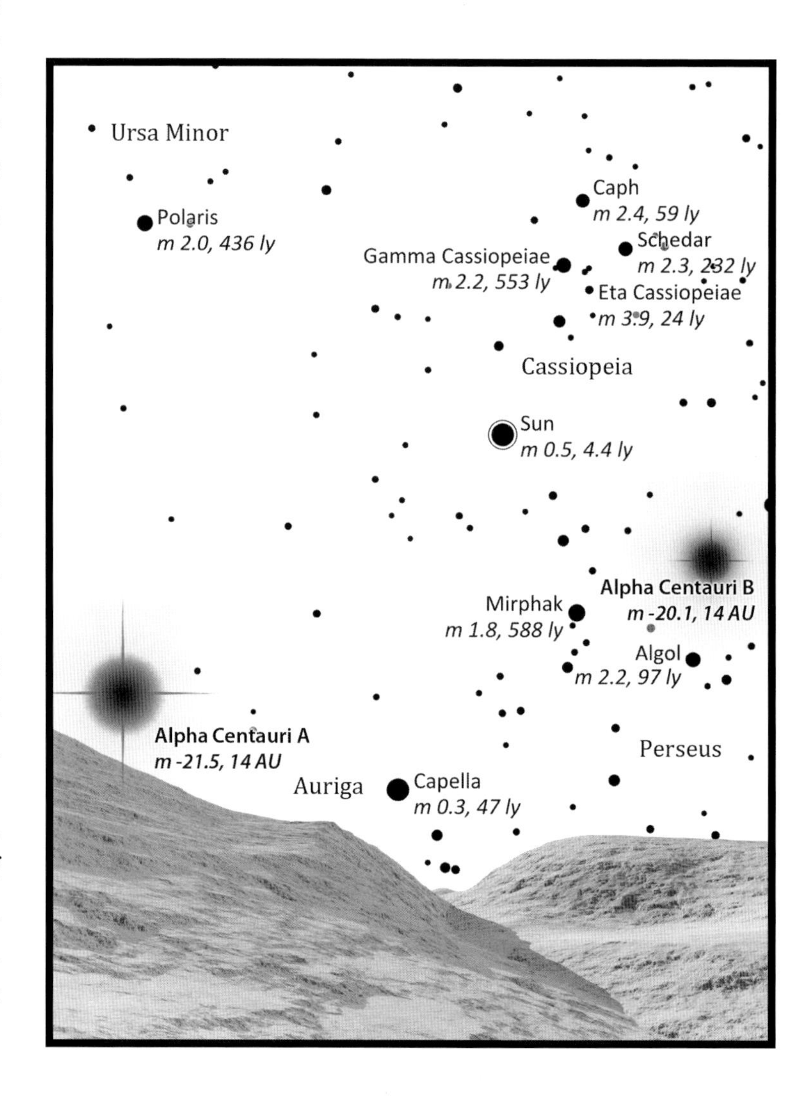

SIRIUS

The Dog Star and its Dwarf Companion

As we tiptoe further out into deep space, we stop by the environs of Sirius—currently the brightest star in the Earth's sky. Sirius is twice as massive as the Sun, much hotter, and 25 times more luminous. It is nearly double the distance of Alpha Centauri, at 8.6 light years from the Sun.

Like Alpha Centauri, Sirius is a double star with individual components (A and B) that harbor no known planets. Unlike the Alpha Centauri pair, the components of Sirius are radically dissimilar. Sirius B is merely a white dwarf—an Earth-sized core remnant of a once massive star that expired 120 million years ago. The material in B is compressed so densely that one teaspoon of it would weigh fifteen tons on Earth. B still emits an intense light of its own, but its diminutive size makes its overall luminosity 8,000 times fainter than A.

From a distance of 1 AU, the brilliant Sirius A floods our camera with a slightly blue-tinged glare, but familiar stars shine through it in the background, as does Sirius B. We may note a trio of other bluish-white stars in our field of view—the famous stars Deneb, Vega, and Altair, which comprise a popular northern hemisphere asterism called "The Summer Triangle." Compared to how they look from Earth, however, Vega and Altair seem out of place. Both have scooted a few degrees toward the position of the Sun, where it resides conspicuously in Hercules as a 2nd magnitude star.

The age of the Sirius system is estimated to be in the range of 225 to 250 million years—a mere child compared to our "middle-aged" Sun's 4.57 billion years. Its Galactic orbit is similar to numerous members of the so-called *Sirius Stream,* which is a group of stars that are infalling together toward the center of the Galaxy. Sirius will come 1,000 light years closer to the Galactic center before swinging back out to a maximum distance of 33,000 light years.

Sirius' supreme brightness in the Earth's sky has transfixed the human imagination since ancient times. It is also the subject of a centuries-old controversy regarding a possible change in the star's color. A number of venerable authors of antiquity—Homer, Aratus, Cicero, Horace, Seneca, and Ptolemy—have described Sirius either as resembling copper or as reddish. Seneca explicitly described Sirius as being "redder than Mars" and Ptolemy put Sirius in league with Betelgeuse, Antares, Aldebaran, Arcturus, and Pollux as examples of "fiery red" stars. Writings from the 10th century onward however, make no such comparisons. Up to the late 19th century, the testimony supporting a historical change in the star's appearance seemed impressive. The advent of 20th century astrophysics, however, threw the matter back into doubt. Such radical stellar evolution on a timescale of just 1,000 years was deemed impossible. Moreover, scholars began to realize that the ancient testimonies were not as unanimous as they initially seemed. Other ancient Latin sources, Marcus Manilius and Avienus, both described Sirius as "sea-blue" and Chinese records dating from the same period deemed Sirius to be white.

Perhaps the famous poets who described Sirius as "reddish" recalled how it looked low on the horizon when its earliest annual appearance at daybreak traditionally heralded the "dog days" of summer. And perhaps other ancient sources were content with parroting fashionable idioms in lieu of making their own observations. But other stars are indeed copper-like and fiery. We shall visit two such stars next.

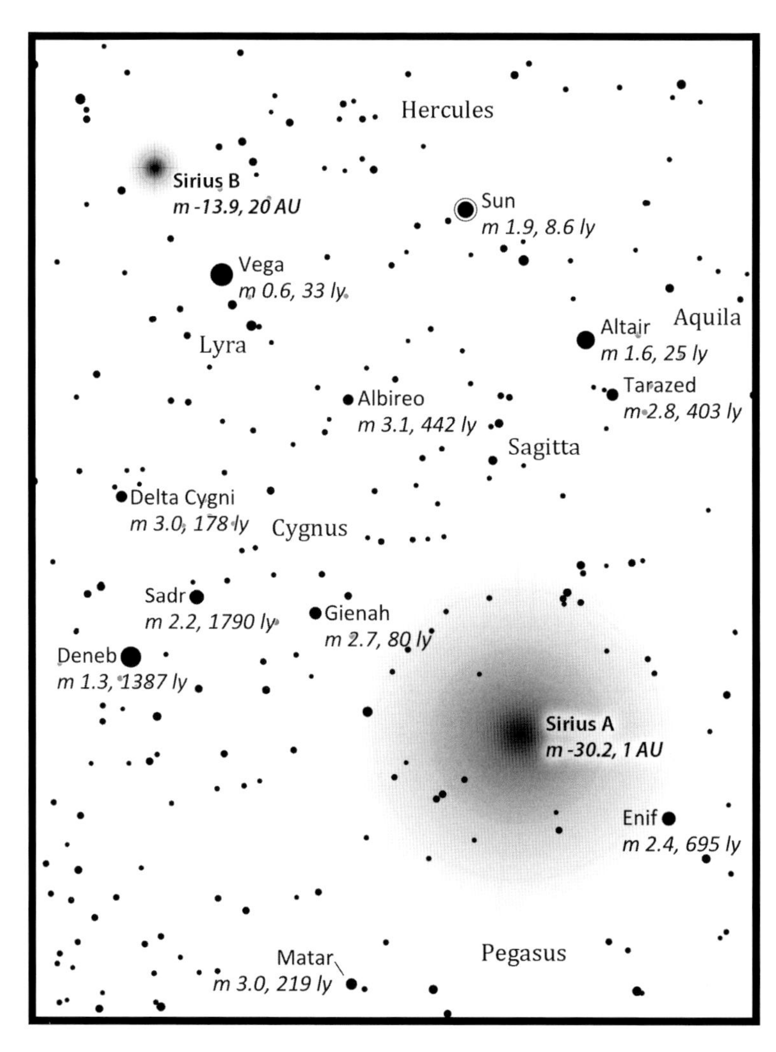

Epsilon Eridani

Planetary System in the Making

We now visit Epsilon Eridani—10.5 light years from the Sun. The star is slightly smaller (three-quarters as big) and somewhat less luminous (one-third as bright) than the Sun and casts an orange-tinged sheen.

From up close, we find its most conspicuous feature to be the circumstellar disk that surrounds the star in a way that is reminiscent of Saturn's rings. Two distinct rings are visible. The inner ring has a radius of 3 AU (275 million miles)—comparable in size to the solar system's own asteroid belt. It barely protrudes from the star's glare from our vantage point above the outer ring, which resides 20 AU (1.85 billion miles) out from the star. Both are comprised of materials that may range in size from fine-grained dust to large asteroids.

Disks of this type may also be termed *protoplanetary disks* because astronomers consider them to be the progenitors of planetary systems. Hence, the environs of Epsilon Eridani perhaps resembles our own solar system in its early, formative years. Our Sun too was once surrounded by a protoplanetary disk at the time of its birth. Over time, the material in the disk assimilated into increasingly larger bodies. These bodies had grown into planets by the time the material was exhausted.

Astronomers have long hoped to find planets hosted by Epsilon Eridani. The broad gap in the disk suggests the presence of massive planets. But reports of a direct detection of even one massive planet are contradictory and remain inconclusive. The likelihood of finding large planets continues to diminish as astronomers scrutinize the system with advancing observational capabilities.

Protoplanetary disks may be a common feature among newer stars (Epsilon Eridani is just several hundred million years old). Only the disks of nearby stars are easy to detect—so far. Epsilon Eridani is also known to possess a third debris disk of icy material (not visible in this portrait) analogous to the solar system's Kuiper Belt. It stretches from 35–90 AU (3.25–8.35 billion miles).

In the background, we may again notice that some prominent stars have shifted position from the places we are accustomed to finding them in the skies of Earth. Alpha Centauri, for instance, has taken up residence next to Zubeneschamali—the brightest star in Libra. Vega has scooted over from Lyra into Hercules. Their displacements are directed toward the Sun's location, found in the constellation Serpens Caput.

Epsilon Eridani was one of two selected targets in a famous 1960 radio astronomy experiment to search for artificial radio signals originating from intelligent civilizations. (The other target was Tau Ceti, which we shall visit soon.) This first experiment of its kind was whimsically dubbed "Project Ozma" (named after Princess Ozma, a character in the celebrated stories of Oz) by the young radio astronomer Frank Drake.

The impressionable 29-year-old project leader and his colleagues exploded with excitement when the 26-meter dish antenna at Green Bank, West Virginia picked up a powerful pulsing signal only a few moments after being aimed at Epsilon Eridani. The celebration, however, was short-lived. The signal faded away just as suddenly and the astronomers soon realized that it came from an airplane passing by overhead.

With the wisdom of hindsight, made possible by recent observations, it is apparent that Epsilon Eridani is not a promising locale for an extraterrestrial civilization. But we may still look elsewhere. There is no shortage of places in our Galaxy to search.

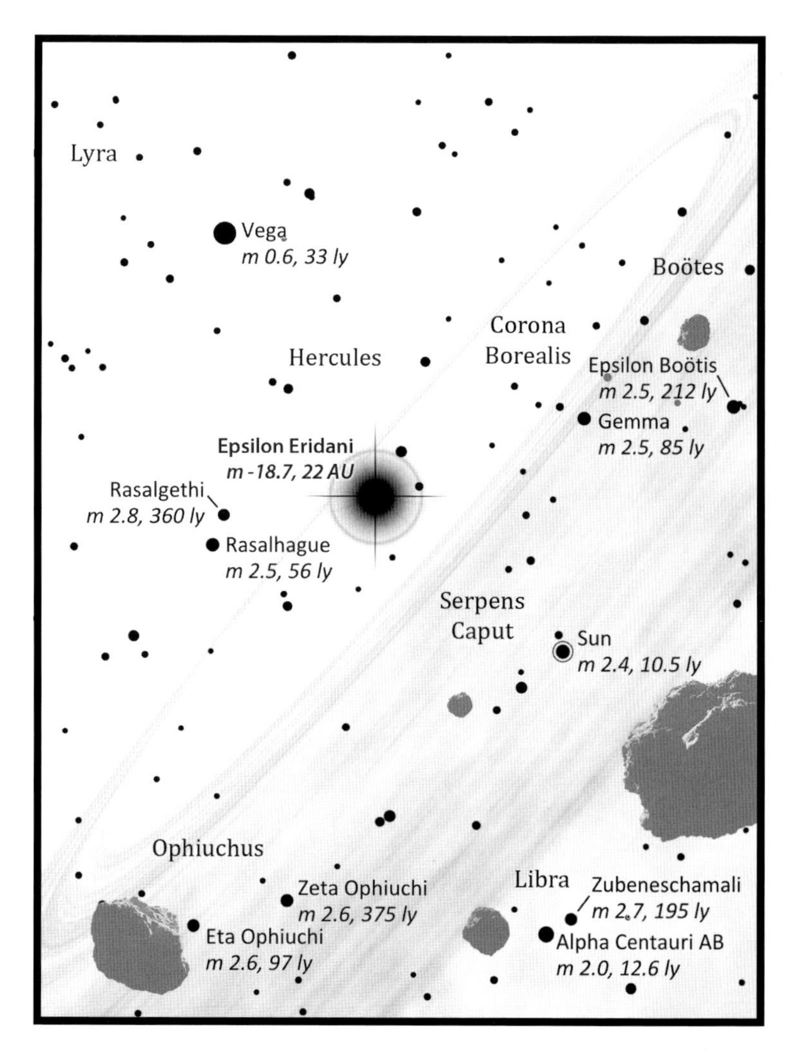

Epsilon Indi

Harbor of Failed Stars

For stargazers in the Earth's southern hemisphere, Epsilon Indi can be seen as a faint orange star 11.8 light years away. But from a distance of 1,460 AU (135 billion miles), the star blooms into a blazing copper torchlight against the jet-black background of space. The star is approximately the same size as the last star we visited, Epsilon Eridani (three-quarters as big as the Sun), but Epsilon Indi is even less luminous (one-eighth as bright as the Sun) and slightly deeper orange in hue.

It is here at the star's periphery that we find a pair of *brown dwarfs*—exotic super-planets that are more massive than the largest gas giants, but not quite massive enough to ignite into stars.

We see one of them, Epsilon Indi Ba, up close. Its twin, Epsilon Indi Bb, is 2.6 AU away from it. It appears as a tiny speck in the distance (see chart). The twins orbit around their common center of gravity (nearly midway between them) every 15 years, and the pair itself slowly orbits their parent star over a period of perhaps thousands of years.

These brown dwarfs are among the closest to our solar system known to date. Each is similar in size to Saturn, yet at least fifty times more massive than Jupiter. It is not yet known how common brown dwarfs may be in the Galaxy. Those that are lurking in interstellar space, unilluminated by parent stars, are much harder to find. Space-borne infrared telescopes (e.g., NASA's Wide-field Infrared Survey Explorer, "WISE") are just beginning to discover them. If brown dwarfs are ubiquitous, one could even be found nearer to our Sun than Alpha Centauri (seen in this portrait to the upper right of Epsilon Indi Ba). But initial results from WISE announced in 2012 concluded that (at least within 26 light years of the Sun) normal stars outnumber brown dwarfs by six to one.

The familiar "Big Dipper" asterism also appears in this portrait, at the center. Aligned with the two stars at the base of the dipper's "bowl" is a third star—our Sun, which appears there at magnitude 2.6.

From the point of view of our Sun, Epsilon Indi has the fastest space motion of any star that we've visited so far. Its remarkable proper motion (the angular speed across the Earth's sky) was noticed by astronomers as far back as the mid-nineteenth century. It moves 4.7 arcseconds per year, or, to put it another way: in 383 years the star transverses a section of the sky equivalent to the width of the full Moon.

Epsilon Indi's velocity through space relative to the Sun clocks in at 90 km/s or 200,000 mph. This is a consequence of the configuration of its Galactic orbit. Its orbital eccentricity (a measure of non-circularity) is pronounced. Epsilon Indi's distance from the Galactic center varies between 15,000 and 30,000 light years. Like Alpha Centauri, the system is on the outbound leg of its Galactic orbit, flying out toward its maximum orbital distance. It will arrive to within 10.6 light years of the Sun 17,400 years from now before receding.

Epsilon Indi, like other stars near the Sun, recurs often in science fiction stories. Perhaps the most memorable character to hail from this locale was Gorgan, the "Friendly Angel" in the televised *Star Trek* episode "All the Children Shall Lead." The character was played by the famed California attorney Melvin Belli. Gorgan turned out to be not friendly at all, but utterly evil. It is perhaps fitting that the actual worlds of Epsilon Indi discovered so far are hellish and dark.

Let us trek on.

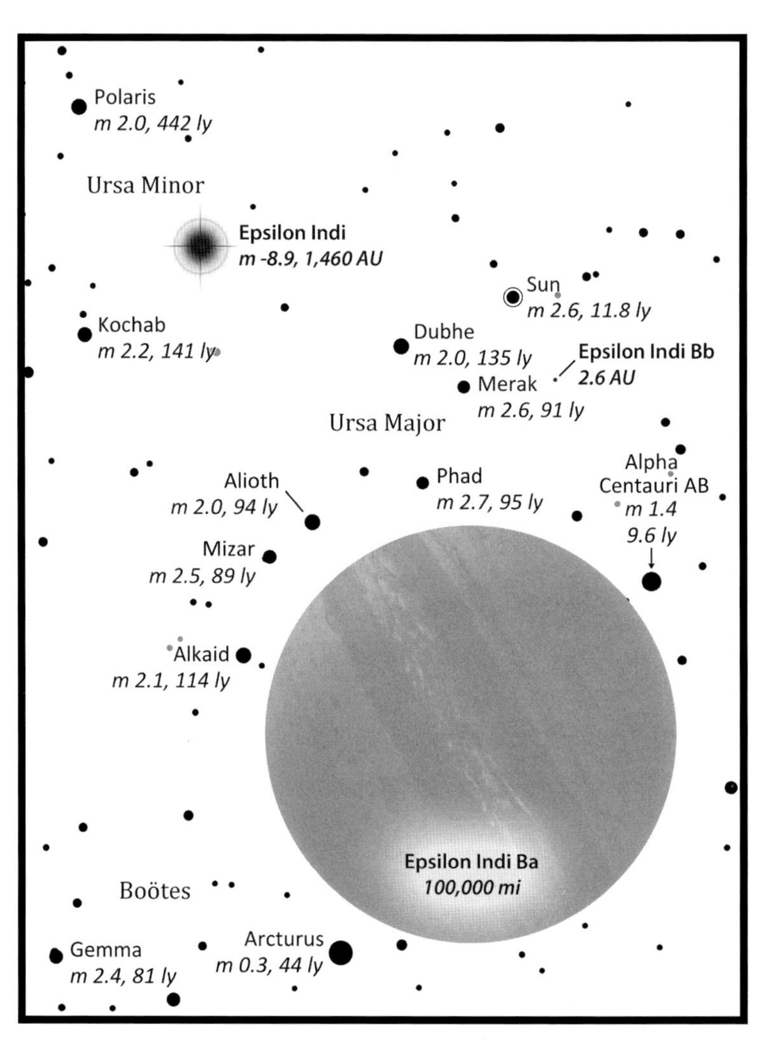

Tau Ceti

Carousel of Comets

Our next stop is Tau Ceti, located 11.9 light years away from the Sun in the constellation Cetus the Whale. Tau Ceti is more Sun-like than the previous two stars we visited, having a size and luminosity comparable to Alpha Centauri B.

Tau Ceti has a broad circumstellar disk, which extends from at least 10 AU to 55 AU. It is likely comprised of a mixture of dusty and icy materials. Tau Ceti's disk is ten times as dense as the solar system's Kuiper Belt, which came as a surprise to the astronomers who discovered it.

Pinpointing the age of Tau Ceti is elusive. It may be just 300 million years old or any age up to eight billion years old! An advanced age for Tau Ceti poses a curious problem: how is it that so much debris remains after all this time? Possibly it is because Tau Ceti, unlike the solar system, does not host any massive planets that would have helped mop up the debris. But then again, we cannot yet say whether the dense debris disk of Tau Ceti is unusual for a mature single star, or if it is the much sparser disk of our solar system that is unusual. We'll have to wait for observational technologies to improve before astronomers on Earth can conduct a deep enough census. A young age for Tau Ceti poses other problems. Tau Ceti possesses a significantly smaller proportion of heavy elements than the Sun (only 35% as much) and it moves in a highly eccentric orbit—parameters that typify the Galaxy's more ancient stars.

Tau Ceti's Galactic orbit has the largest eccentricity of any of the stars we've visited so far. Its distance from the Galactic center varies from a minimum of roughly 24,000 light years (near the present position of the Sun) to a projected maximum of 72,000 light years—perhaps beyond the farthest traces of the Galaxy's spiral arms. It is presently on its inbound leg, overtaking the Sun at a relative velocity of 37 km/s (83,000 mph). It will pass by the Sun 43,000 years from now at a distance of 10.6 light years before wayfaring back to the edge of the Galaxy.

Tau Ceti's location in the skies of Earth is within 17 degrees of the South Galactic Pole—one of two points exactly perpendicular to the plane of the Milky Way at the Sun's location. Looking back at the Sun from Tau Ceti, we find it within 17 degrees of the North Galactic Pole. Consequently, the star field is only sparsely populated with bright stars. Arcturus happens to be nearby, as is the Coma Berenices star cluster.

As we mentioned while visiting Epsilon Eridani, Tau Ceti was the second of two targets in Frank Drake's "Project Ozma." No alien transmissions were detected. As in the case of Epsilon Eridani, the wisdom of hindsight informs us that Tau Ceti, too, was not a promising target. A terrestrial planet within 1 AU of the star, the vantage point of this portrait, would be inundated with comets from Tau Ceti's densely populated icy debris disk. It might be impacted ten times more often than the Earth is in our solar system. That wouldn't be a propitious environment for the evolution of complex life.

While no planets have yet been detected around Tau Ceti, it is possible that its icy debris disk hosts many Pluto-like worlds analogous to the solar system's largest Kuiper Belt objects. For all we know, some could even be as large as the Earth. But their large orbital distances would make them too frigid for life as we know it.

Let us next visit a star that is now well-known to possess actual planets.

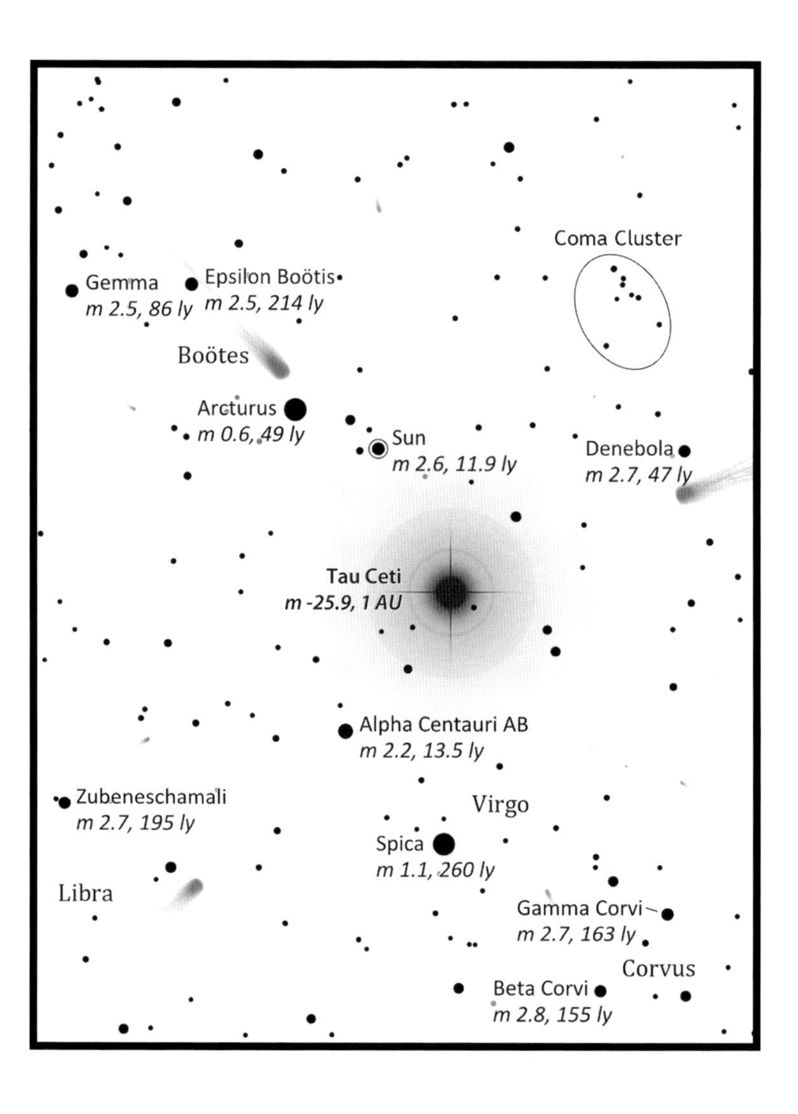

Gliese 876

A Planetary Quartet Plays the Music of the Spheres

Gliese 876 is a red dwarf star approximately one-third the size of the Sun yet six hundred times less luminous in terms of visible light—twice as faint as the white dwarf Sirius B. A large pair of binoculars or a small telescope is required to see it from Earth, even though it is just 15.3 light years away. Also known as "Ross 780," it can be found as a magnitude 10.2 star floating in the "Water Jar" of the constellation Aquarius.

Closing in to a distance of 0.4 AU (37 million miles) we find the star perked up to a respectable brilliance. Also within view are its four known planets, all of which are much more massive than the Earth. The smallest and innermost planet (Gliese 876 d) is at least six times more massive than the Earth and the largest (Gliese 876 b) is at least twice the mass of Jupiter.

The orbits of the three outermost planets (e, b, and c) have settled into a *harmonic resonance*. Their orbital periods are fixed at a 1:2:4 ratio—that is, every time planet e completes one orbit (once every 124¼ Earth days), planet b will have completed two orbits and planet c will have completed four orbits. A similar example can be found in our own solar system: three of Jupiter's Galilean moons, Ganymede, Europa, and Io, are choreographed to an identical rhythm. Unlike the Galilean moons, however, Gliese 876's planets also have an element of synchronicity in their dance, so that e, b, and c nearly form up in a straight row whenever e reaches a key position in its orbit—a "triple conjunction." The innermost planet, d, however, dances to a more mercurial tempo. It whips around Gliese 876 once every two days. But it would still be possible to witness a quadruple conjunction if one chooses the right moment.

We capture the moment in this portrait by opting for a super-telephoto perspective (unlike the wide angle aspects of all of our previous destinations), which gives us a magnified view.

We take up a vantage point far behind e and line up our snapshot as the other planets file into place. We find the large planet b in the center of the image, lighting up the dark side of c, above, with its reflected light. We see the inner planet d transiting across the active chromosphere of Gliese 876, which is roiling with large spots and prominences—signs of its youth.

We may also infer Gliese 876's young age because it is co-moving with other stars in the Beta Pictoris *moving group*. These stars are *coeval* (meaning "born together") and range between 10 to 30 million years old. Other members of this group are known to have circumstellar debris disks. Their Galactic orbits, also typical of young stars, follow fairly round paths (i.e., low eccentricity). Gliese 876's position near the Sun is also its maximum distance from the Galactic center—24,000 light years. It is just beginning to fall back to its minimum "Galactocentric" distance of 22,000 light years.

The extreme zoom employed in this image has spread the background stars quite thin. The familiar constellations are too large to fit in the narrow field of view. The Sun lies outside this portrait, but is otherwise easily locatable as a magnitude 3.2 star midway between Regulus and Denebola in the constellation Leo.

Let us now continue venturing further away from our homeworld.

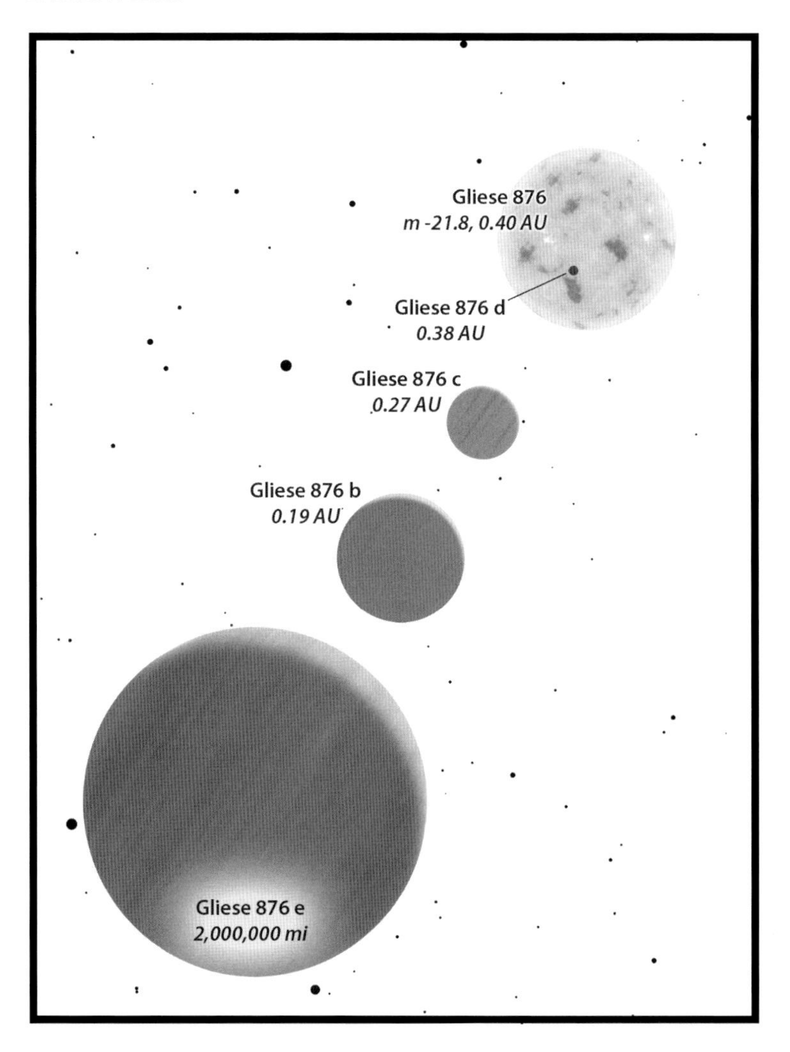

ALTAIR

Foreboding Star, Forbidden Planet

Our next subject is Altair, the hottest and most luminous star that we've visited since Sirius. It shines at magnitude 0.8 in the skies of Earth—it is the "eye of the eagle" in the constellation Aquila.

For our close-up visit to Altair, we resume a wide-angle view, which frames the dazzling white orb in a rich backdrop of bright stars. Orion is conspicuous to the lower right. Sirius and Procyon, the principal stars of the constellations Canis Major and Canis Minor, have converged toward the Sun's position in Monoceros. Three other stars we've visited—Alpha Centauri, Epsilon Eridani, and Tau Ceti—can also be found in the field.

Our camera is positioned just 0.05 AU (5 million miles) away from Altair so that we may discern its remarkable characteristics. Altair spins rapidly—one rotation every nine hours—which is slightly faster than Jupiter's spin rate (10 hours) and much faster than that of the Sun (25 days). Its rotational velocity at the equator is estimated to be in the range of 285 km/s (640,000 mph). This sets up a powerful centrifugal force that causes Altair to bulge out at the equator. The resulting shape is an oblate spheroid that spans twice the Sun's diameter from side to side but only 1.6 times the Sun's diameter from top to bottom. Another peculiar consequence of Altair's rapid rotation is that its surface gravity is weaker at the equator, which reduces surface temperatures in the equatorial region. These lower temperatures make the equator slightly darker than the polar regions.

Altair's proximity to the Sun, just 16.7 light years away, has often served to inspire science fiction creators. The storyline of the groundbreaking 1956 sci-fi film *Forbidden Planet* took place on the imaginary planet "Altair IV." The classic screenplay tells of a lost alien civilization, the Krell, that died out hundreds of millennia before space-faring explorers from Earth first arrived. The initial colonization effort did not go well; all but two survivors succumbed to the Krells' mysteriously lethal legacy. Subsequent visitors would only be able to survive by solving the mystery.

Futuristic fables notwithstanding, Altair itself poses real natural dangers. Any possible planet suitable for human habitation would have to orbit Altair at a safe distance. Altair radiates eleven times as much energy into space than does our Sun. A world located at 3.3 times the Earth-Sun distance would be bathed in Altair's light at energy levels comparable to sunlight on Earth, but would also receive a higher proportion of it as ultraviolet radiation.

Altair is estimated to be 1.3 billion years old and is twice as massive as the Sun. But because it burns through its hydrogen fuel more quickly, it has already surpassed the midpoint of its stellar lifetime. In another one billion years, all its fuel will be exhausted and the star will begin to expire.

Altair's Galactic orbit has a lower eccentricity than the Sun's. This means the orbit is more circular, as is typical of stars that are younger than the Sun. It is presently journeying outward to its slightly further maximum distance from the Galactic center of 27,500 light years. The Sun is overtaking Altair and will close to within 8.6 light years of it 138,000 years from now. In that span of time, Altair's brilliance in the Earth's skies will increase fourfold to magnitude -0.7.

Altair is even more foreboding than the creators of *Forbidden Planet* ever imagined. But other nearby stars may be as amenable as our familiar Sun. Let us visit two of them.

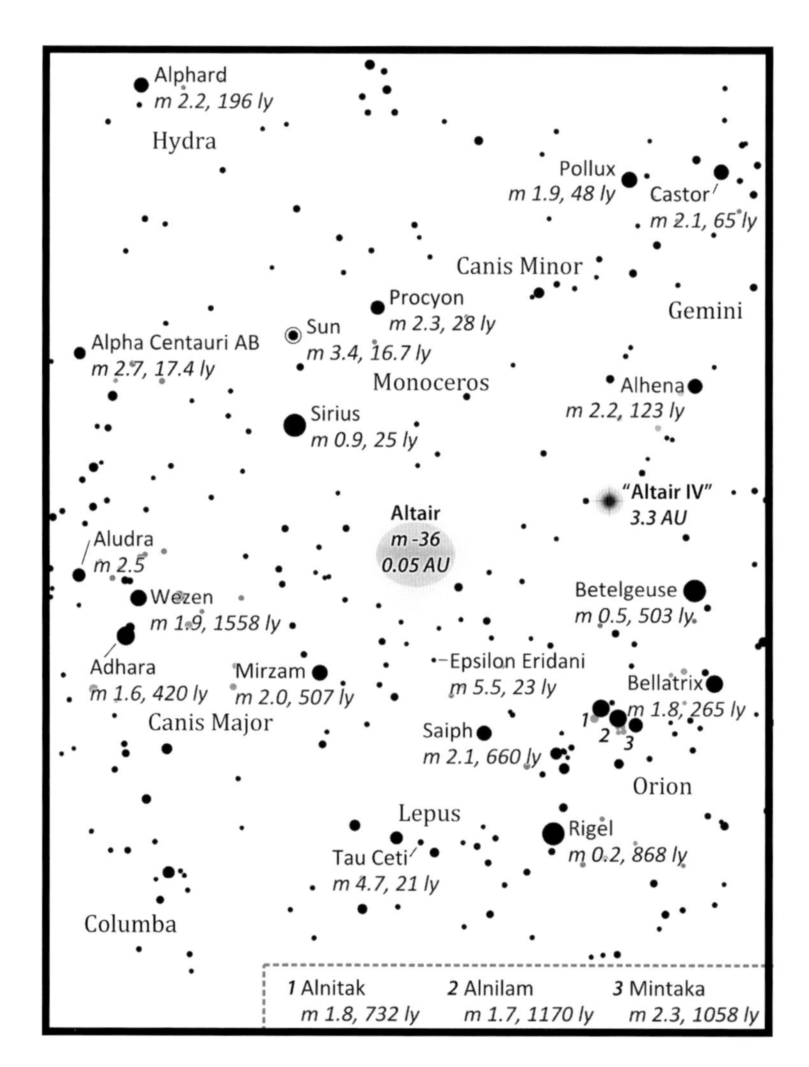

Eta Cassiopeiae

To Sail Beyond Sunset

Eta Cassiopeiae is a double star system located 19.4 light years away from Earth. The primary star, Eta Cassiopeiae A, is comparable to the Sun—virtually identical in size and similar in color. Its chemical content, however, is somewhat less abundant in heavy elements. It burns 30% brighter than our Sun even though it is only 95% as massive—a sign of advanced age, which is estimated (with much uncertainty) to exceed 5 billion years.

The fainter secondary star, Eta Cassiopeiae B, is a red dwarf, discovered by the world-renowned astronomer William Herschel in 1779. The pair exhibits an appealing contrast of colors. B orbits A in a highly elliptical trajectory that varies in distance between 36 AU (slightly more than the distance between the Sun and Neptune) and 107 AU.

Here we visit a hypothetical Earth-sized planet in orbit around the Sun-like star A. The planet is covered by a vast ocean. The weather happens to be calm on a cloudless evening. Because it is nighttime, A is not visible, but the ruddy light of far-away B bathes the watery world with the brightness of three full moons.

It is thus fitting that the constellations crowding the skies above are some of the ones visible from the south seas of Earth. The star charts inherited from classical times by 16th century explorers were devoid of southern stars—a celestial *terra incognita.* That void was filled in with "modern" constellations, many of which adhere to a nautical theme. Hence we find Carina the Keel and Vela the Sails at the right edge of this portrait. European sailors left the southern skies littered with other maritime gear: a compass and an octant are also included among the southern constellations. But surely the most iconic constellation of the southern skies is Crux—the Southern Cross. Crux can be found in the upper center of the portrait. Tucked into a corner of the cross we find the Sun and Alpha Centauri as 3rd magnitude stars.

What might future generations of explorers discover on such a world as this? No continents are to be found here. Perhaps this planet lacks the geological dynamism of plate tectonics, so that no large expanses of land have ever been uplifted above the surface of the sea. Or perhaps it was deluged by too many comets during the epoch of its formation, when we might suppose that Eta Cassiopeiae's reserve of Kuiper Belt-like objects has already been dispersed to exhaustion by the relentless stirring of the widely wandering B.

Even so, this world may be dotted by smatterings of small volcanic islands. Perhaps one such oasis lies just over the horizon, waiting to be developed into the ultimate getaway: an extraterrestrial Tahiti, where space explorers may retire to tan themselves by the rays of an alien star and lazily catch up with latest news transmissions from Earth—twenty years after the reported events happened.

Curiously, Eta Cassiopeiae follows a Galactic orbit that closely parallels the previous star we visited—the 1.3 billion-year-old Altair. The relationship must be purely coincidental. The stars certainly did not have a common origin. But no matter the circumstances that married them, their close proximity to each other may endure for hundreds of millions more years. These perennial traveling partners may even continue their Galactic cruise together until the very day that the faster-aging Altair dies off.

For our own part, we shall cruise onward to one last nearby star on this first itinerary.

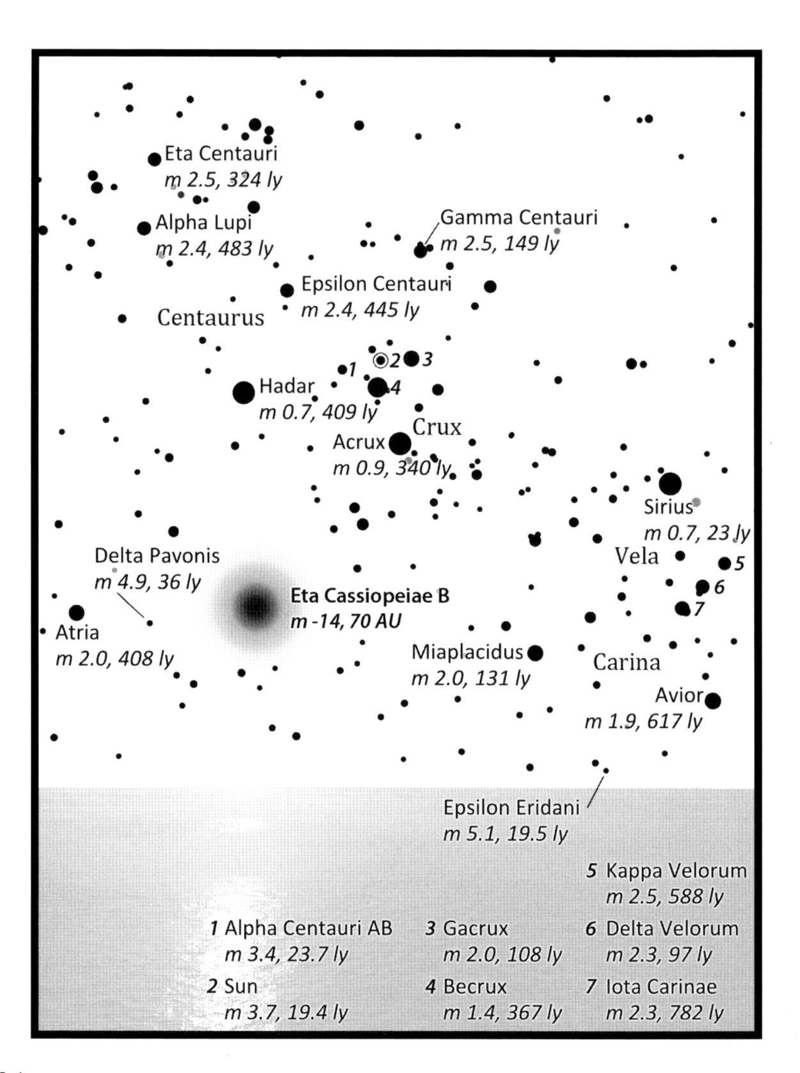

Delta Pavonis

An Old Man of the Galaxy Ages Gracefully

Delta Pavonis is a single star nearly 20 light years distant from Earth in the southern constellation Pavo the Peacock. Its mass is very nearly equivalent to the Sun's, but it burns slightly brighter and cooler. It is classified as a *subgiant* rather than a *main sequence* star, which simply means that it is brighter than the bulk of similar stars in its temperature range. This happens either when a star is highly abundant with heavy elements, or because it is running out of hydrogen fuel—a sure sign of senescence. In the case of Delta Pavonis, both reasons hold true. The star is indeed rich in heavy elements and is also estimated to exceed 9 billion years in age, making it quite possibly the oldest star we've visited so far.

Stars that are *metal rich* are believed to be more likely to host planets, or at least potentially more interesting ones, such as complex terrestrial worlds.

No planets of any kind have been discovered in the environs of Delta Pavonis so far. But in this portrait we record another visit to a hypothetical world.

We behold an aerial view. It is dawn, and the light from Delta Pavonis (beyond the right edge of the portrait) filters in across a distant horizon. An icy promontory stands at the inlet of a windswept sea. This planetary environment is in the process of sudden release from a perpetual ice age as the hydrogen-depleted Delta Pavonis ripens in brightness. The ice is thinning, the rising sea level floods the shores, and photosynthesizing organisms are proliferating—enriching the atmosphere with oxygen. Perhaps this warming world is on the very brink of readiness for human inhabitation and may remain so for millions of years until Delta Pavonis eventually balloons into an intolerable giant. In the meantime, the old star ages gracefully.

In Frank Herbert's science fiction series *Dune,* Delta Pavonis is the parent star of the oceanic homeworld of the Atreides dynasty, the planet Caladan. The skies of Caladan would not look too much different to the young Paul Atreides in "the year 10,191" than they appear here in the present. We find the bright stars Sirius, Procyon, Pollux, and Castor grouped tightly together into an keystone-shaped asterism. We also find Capella nearby, still in the constellation of Auriga, though the outlines of the constellation are distorted almost beyond recognition. Consulting the sky chart, we find nearly all of our previous destinations arrayed before us. Only Altair and Gliese 876, which lie in other directions, are not contained in this image. The brightest star that appears from Delta Pavonis, also beyond this frame's borders, is Canopus—designated by Herbert as the parent star of the main locale of the series, the desert planet Dune.

Delta Pavonis, like our Sun, follows a moderately eccentric Galactic orbit, but is presently outbound—on its way to its maximum Galactocentric distance of 29,000 light years. Given Delta Pavonis' advanced age, this may be its final lap around the Galaxy as a normal star.

For our own part, we may now finish our short lap around the solar neighborhood by returning home. We shall remain there for the duration of our next itinerary, which won't be a voyage through space, but a long journey through time.

But before we get started, let us take an overview of the places we have visited so far.

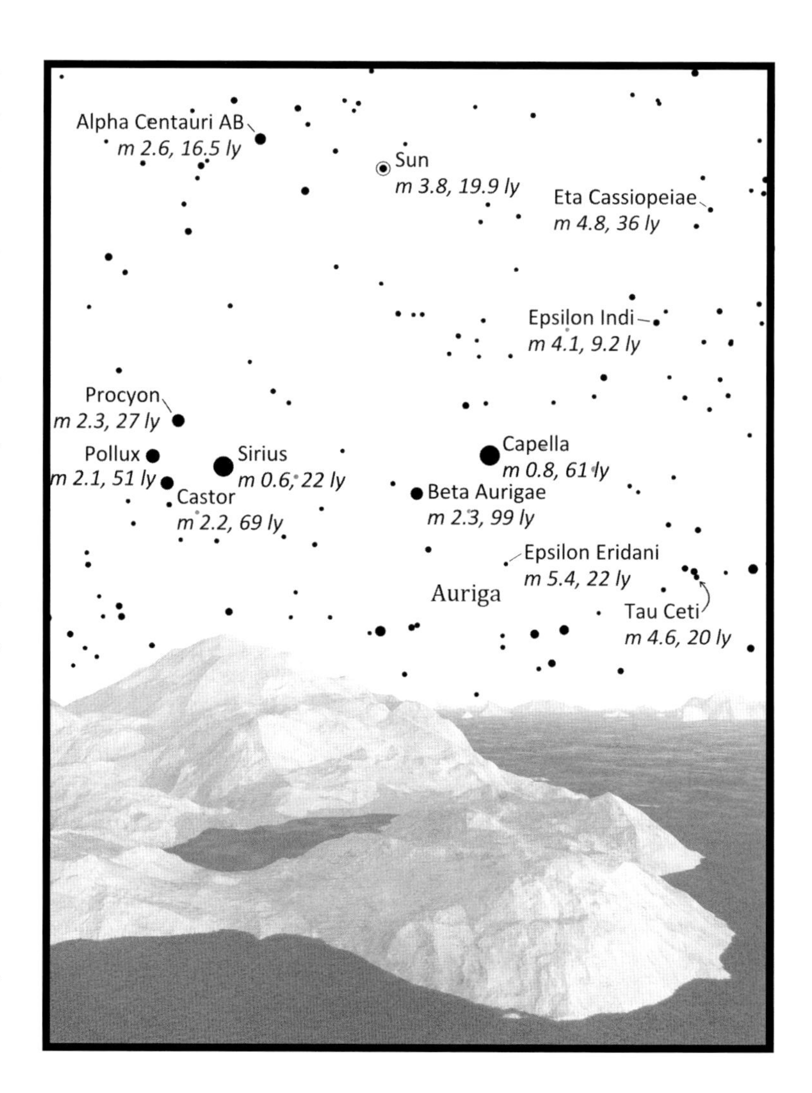

Overview

The Motions of Stars Around the Galaxy

"I have found it very probable that the so-called fixed or firm stars could really be slowly moving—wandering stars of a higher order."

- Immanuel Kant, 1755

We have visited ten stars presently within twenty light years of the Sun. Along the way, we occasionally noted some characteristics of their Galactic orbits. The illustration on the opposite page gives us the big picture: the orbits of all ten stars. A word of caution: Galactic orbits are not really this neat. The gravity exerted by the total mass of the Galaxy imposes deviations. Consequently, actual Galactic orbits do not exactly follow these idealized closed ellipses. But the diagram does give a fair indication of how near and how far these stars wander to and away from the Galactic center.

As the solar system completes each Galactic orbit every few hundred million years, the other stars that it encounters along the way make guest appearances in the skies of Earth. One Galactic orbit ago, an entirely different retinue of stellar guests had adorned the Earth's celestial vault, and yet another retinue will be in place one Galactic orbit hence.

The limits to measurement precision of the Hipparcos mission (page 8) allow us to project the exact courses of stars for only a few million years into the past and future with great confidence. But even in this limited range of time, all of the constellations now familiar to us have undergone and will undergo radical transfigurations. The sequence of charts to the right, for example, document the ongoing evolution of the Big Dipper asterism. Over the next 150,000 years, its shape will invert: the stars of the handle will turn into a bowl, and the stars of the bowl will turn into a handle.

As these and other stars have shuffled around the heavens, so too has the Earth below undergone myriads of changes. Periods of glaciation transformed the landscape in times past, as they will again in the future. Many new species of plants and animals evolved into existence while others died out. The first human beings awoke to experience it all, and the last humans may one day leave the Earth behind to take up residence amidst the stars.

So let us now begin our adventure in time travel, and follow the course of Earth's natural history for ten million years.

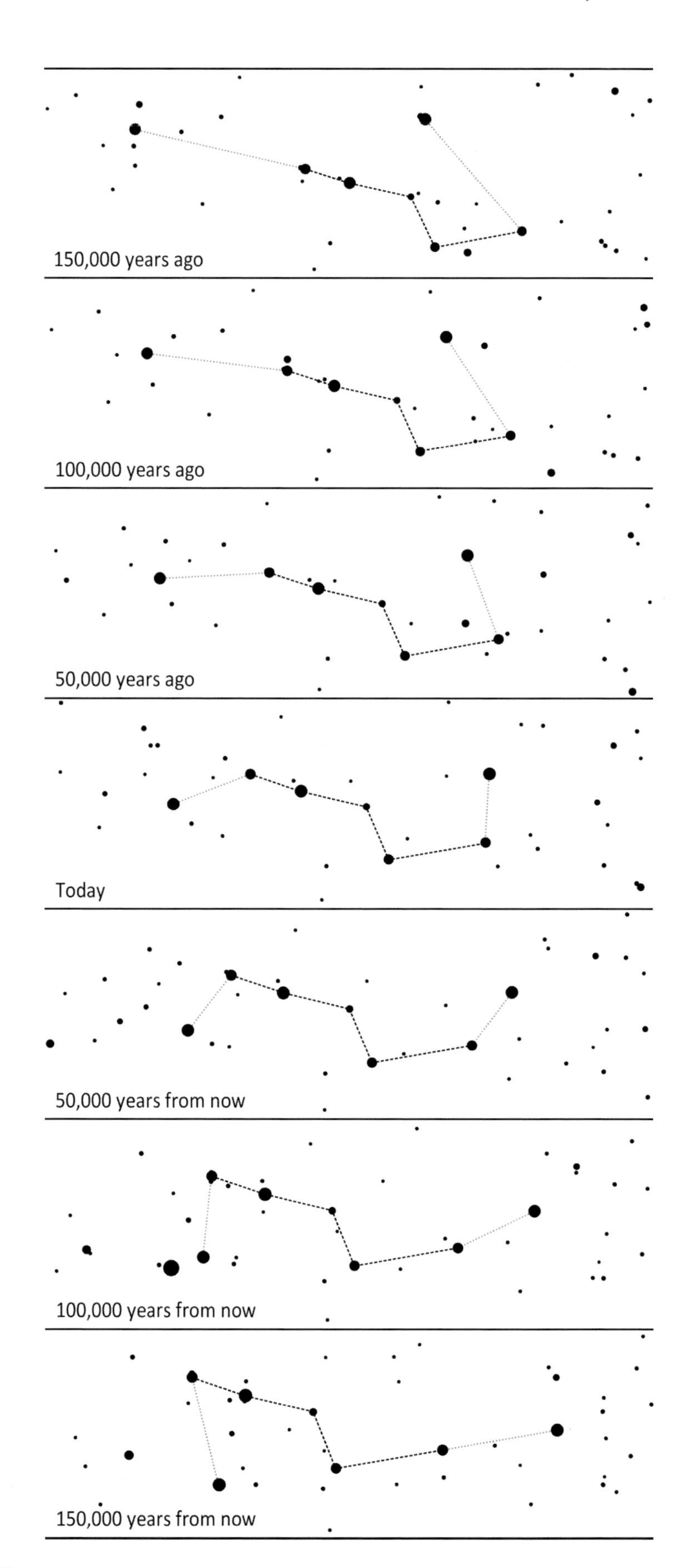

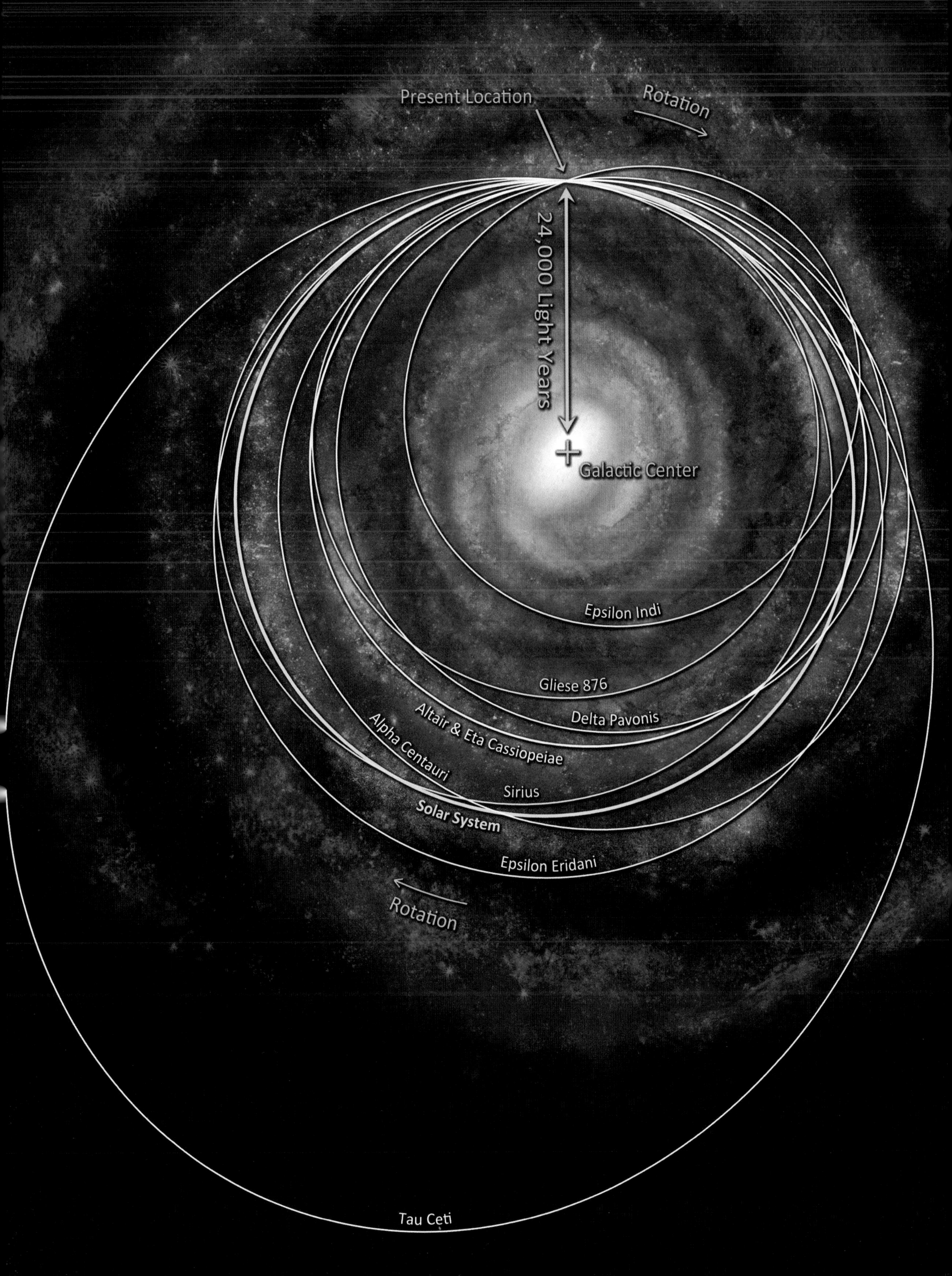

Present Location
Rotation
24,000 Light Years
Galactic Center
Epsilon Indi
Gliese 876
Delta Pavonis
Altair & Eta Cassiopeiae
Alpha Centauri
Sirius
Solar System
Epsilon Eridani
Rotation
Tau Ceti

SECOND ITINERARY

Ten Million Years Aboard Spaceship Earth

The Pillars of Hercules

5.3 Million Years Ago

Once upon a time, late in the Miocene epoch, an enormous chasm straddled the continents of Europe and Africa, where there had previously existed a broad sea. Cut off from the Atlantic Ocean in an era of diminishing rainfall, the sea's inflowing rivers could not replenish its waters fast enough to prevent it from evaporating almost completely. The depths of the resulting desert stooped as far as two or three miles below sea level. There, the only remnants of the once grand sea were shallow salt lakes.

After languishing for a few hundred thousand years in this condition, ongoing geological processes would conspire to allow the basin to be refilled. The Atlantic barrier cracked, a channel opened up, and the world's mightiest river was forged to deliver a thousand cubic miles of cascading seawater each year. Thus the Mediterranean Sea was restored to glory.

This story, parlayed by modern geologists, uncannily resembles an ancient Greek fable. In the myth of the Twelve Labors of Hercules, one of the hero's assignments was to rustle cattle from the island of Erytheia, far to the West, and to herd them back to Greece. In one version of the story, Hercules took a shortcut to his destination by smashing through a daunting mountain. In doing so, he connected the Mediterranean Sea to the Atlantic Ocean, thus creating the Strait of Gibraltar. The two opposing peaks on each side of the Strait have hence been known as the Pillars of Hercules.

In tribute to this ancient tale, we gaze westward toward a region in the sky that will contain the constellation Hercules in our time. We have traveled back 5.3 million years and the Mediterranean Sea is refilling before our eyes. A tremendous moonlit waterfall gushes forth with seawater—the first cascade of a miles-long descent. The apparition of a *moonbow*—a rainbow produced by the moonlight—arcs through the mist. The stars above convey no familiar constellations. Let us examine them.

The Hercules region contains the so-called *solar apex*—a special position in the sky. Most stars appear to "radiate" from there as the Sun orbits the Galaxy. It is located near (but not precisely in) the direction of Galactic rotation. This point is analogous to the center-point of a fictional starship's viewscreen (think of *Star Trek*) as it flies through the Galaxy faster than lightspeed. *Spaceship Earth* is carried around the Galaxy by the motion of the Sun at about 225 kilometers per second, or 500,000 miles per hour. That's about 1,300 times slower than lightspeed, but still fast enough transport the solar system across 4,000 light years of Galactic space over the time period of 5.3 million years.

Some stars have kept pace with us during all this time. The position of distant Deneb, seen at the top of the portrait, has barely budged from where we find it in our time. But most stars near the Sun have space velocities that allow them to be overtaken. Polaris, for instance, appears here near the solar apex. As the Sun overtakes Polaris, it will grow in apparent brightness as the star drifts 35 degrees northward to reach its "pole position" of our epoch. Indeed, many of the bright stars in this portrait are, in our time, converging toward the *solar antapex* in Canis Minor—Spaceship Earth's "rear window."

The brightest star seen in this portrait, Theta Columbae, will become spectacularly brighter in another half-million years. Let us now journey forward into the future and have a look.

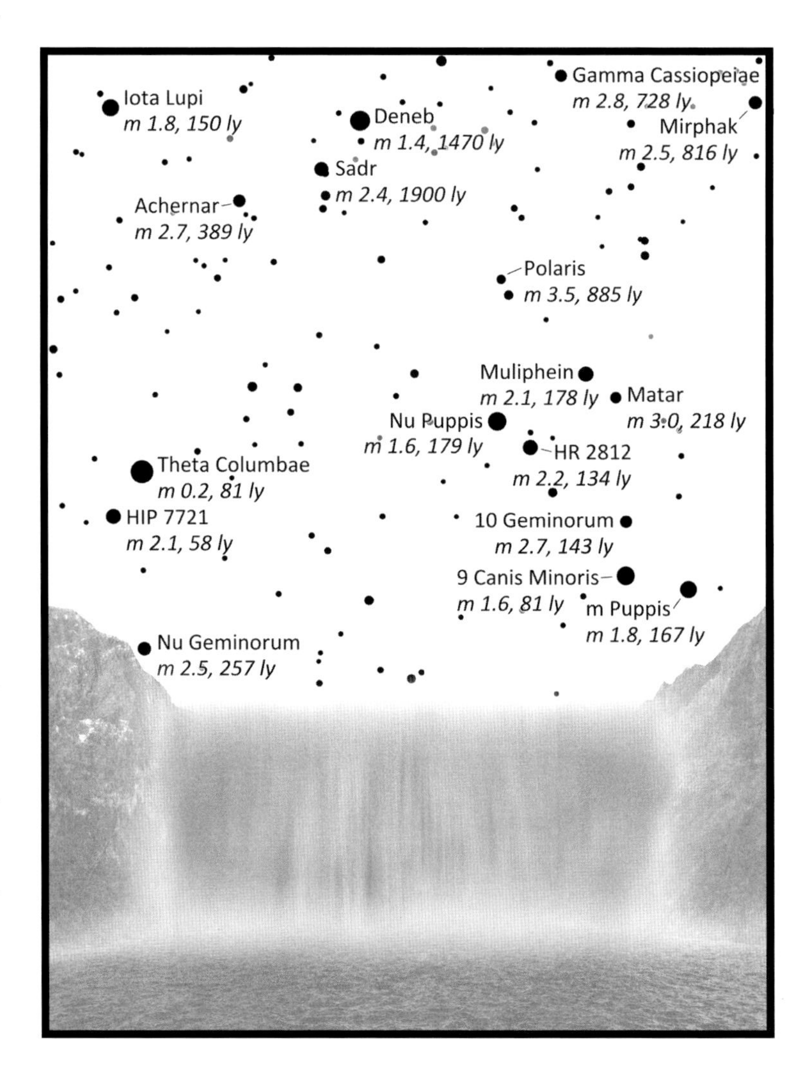

Flyby of a Dove Star

4.8 Million Years Ago

Theta Columbae is, in our time, an unremarkable fifth magnitude star 720 light years away in the southern constellation of Columba the Dove. Being a blue subgiant star more than 700 times as luminous as the Sun, no other star (with the exception of briefly occurring supernovae) may have more conspicuously graced our heavens since it passed within seven light years of the Earth 4.8 million years ago. At that close distance, Theta Columbae was a *negative fifth* magnitude star. This difference of 10 magnitudes equates to a 10,000 times difference in apparent brightness. The star would have even outshone the bright planet Venus. A discerning human eye would have been able to pick it out even in the bright blue skies of daytime—if only there were humans living then.

The keener eyesight of migratory birds may have had an even easier time of it. Perhaps the dazzling Theta Columbae even served as a convenient aid to navigation. How birds navigate is still largely a mystery. Scientific explanations have ascribed their ability to both learning and genetic programming. Possible navigational clues that might be taken from the environment include sounds, smells, magnetic fields—and the stars. Perhaps their amazing ability is derived from combinations of all these clues—and others that we haven't even thought of yet.

In this scene, we observe a flock of an ancient species of geese making its way southward through a mountainous pass somewhere in central Mexico. It is sunset, and Theta Columbae, itself undergoing a migration on a much longer timescale, shines at maximum brilliance in the center of the portrait.

A number of other bright stars also appear here. The giant white star Canopus, 100 light years closer to Earth than it is in our time, shows up at magnitude -1.4. Throughout most of the past several million years, Canopus has been the brightest star seen in the skies of Earth. This is due to its tremendous luminosity (13,000 times more luminous than the Sun) and also because its velocity has coincidentally kept pace with the solar system (within 25 kilometers per second). Only when less luminous stars more closely encounter the Sun does Canopus concede its top rank. In our own time, Canopus has been dethroned by Sirius—but only temporarily. Sirius too will fade into the distance as time slips by and Canopus' preeminence will be restored yet again.

Curiously, we also find the orange giant Kochab deep in the southern sky. In our time, this star is part of Ursa Minor—the northernmost constellation in the modern sky. It is, today, the brightest star in the bowl of the "Little Dipper," just 16 degrees away from Polaris. Kochab belongs to a minority of stars in the solar neighborhood that are outpacing the Sun in the direction of Galactic rotation. Its moderately eccentric Galactic orbit is roughly similar to Epsilon Eridani (see page 14). Kochab is closing in on its minimum distance to the Galactic center, where the star's orbital velocity achieves maximum speed. Hence, its migration runs counter to most other stars—first appearing in the "rear window" of Spaceship Earth as it was catching up to us. In our time, it is moving toward the "front window" as it leaves us behind.

In the parade of stars that march across the skies of Earth at the pace of Galactic time, Theta Columbae's splendor is an extreme example. But two other stars did come close to reaching similar prominence.

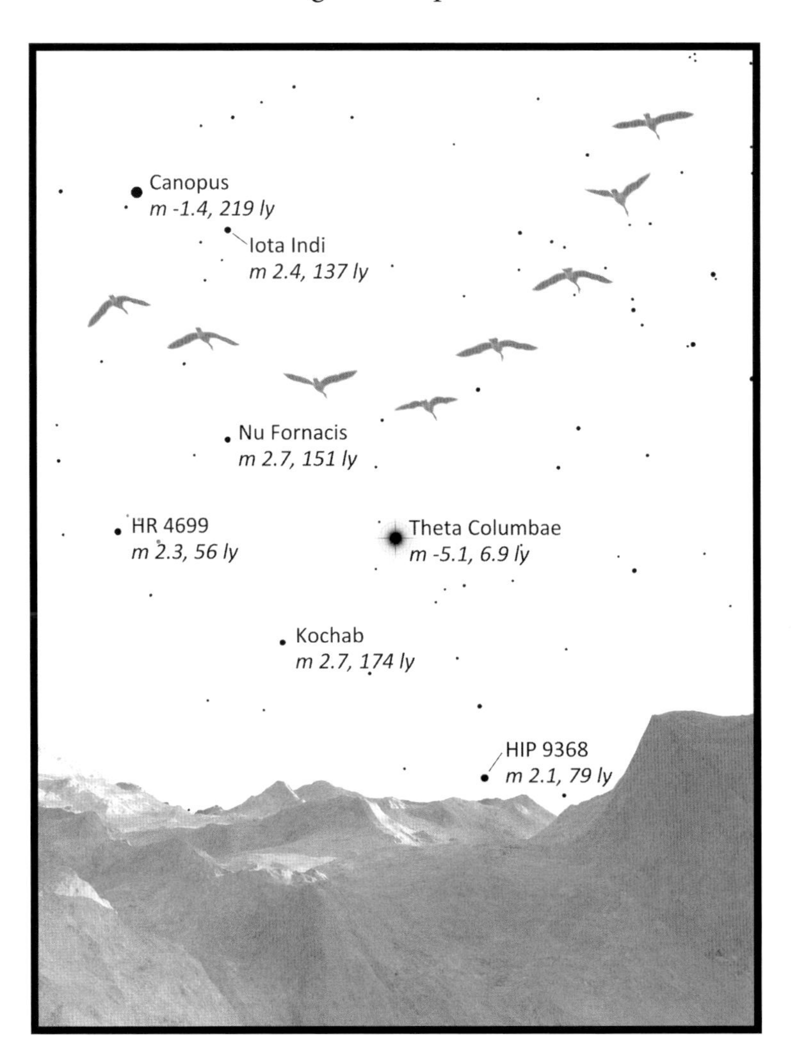

ARDI AND ADHARA

4.4 Million Years Ago

Planet Earth has been a progenitor of living beings for most of its 4.6 billion year history. But only within a tiny fraction of all that time did our planet give rise to human beings. The course of evolution that led to our existence was riddled with fortuitous twists and turns, but also branched off at irregular intervals to hapless dead ends.

We pick up that evolutionary trail 4.4 million years before our time, deep in Ethiopia. This was home to a band of great apes of the genus *Ardipithecus.* We know of their existence through the discovery of fossilized skeletal remains. A female specimen dating from this time is the oldest example of a hominid yet discovered. Her discoverers even gave her a personal name: *Ardi.*

Ardi lived her life in ancient Ethiopia as an ape who could walk upright in her woodland environment. In this scene we find Ardi crouched in a tree at the edge of the forest, overlooking a broad valley at dawn. This was the edge of her universe. The path leading to our human destiny would have to follow a more intrepid hominid branch—a line of apes that ventured onward into the more perilous open savannah.

Above Ardi's arboreal outpost we find a splendid hot blue giant star, Adhara, which glows with great luminosity: 20,000 times brighter than the Sun. We see it here at the time when it happened to have passed closest to the solar system. At a distance of 31 light years, it shines at magnitude -4.1. This is not even half as bright as Theta Columbae had appeared in the previous scene, but it is still 11 times brighter than Sirius appears in the skies of our time.

Adhara also has a twin which flanks it on a near-parallel course through space—another hot blue giant star, Mirzam. Mirzam does not appear in this portrait. It lies behind us—in the exact opposite direction in the sky. It too blazed at similar magnitude (-3.8) at a similar distance (35 light years). Adhara and Mirzam are still conspicuous in our time, in the "feet" of Canis Major, 400 and 500 light years away respectively.

Another bright star peeks through the branches of Ardi's tree: the rapidly-aging massive star Betelgeuse. In our time, Betelgeuse is a cool red supergiant star in the shoulder of Orion, close to the very end of its stellar life. But in the star's younger days, just 4.4 million years before, it had gleamed with a whiter hue.

Can we be confident that this is the location where Betelgeuse appeared in Ardi's sky? Some astronomers suppose that Betelgeuse may have had an even more massive companion that went supernova sometime between Ardi's time and our own. If so, the conservation of orbital momentum of the surviving star—Betelgeuse—would have altered its Galactic trajectory. This change of course would put the star in another place of origin. Though impossible to trace, one plausible alternate location would be closer to where other stars in Orion had originated.

Yet, for all we know, the hypothesized companion may never have existed and Betelgeuse originated from where we can actually trace it. Modern astronomers may continue to debate the issue, but Ardi knew for sure. She saw the whitish Betelgeuse wherever it really was.

Let us now linger in Ethiopia for another million years and see who else comes along.

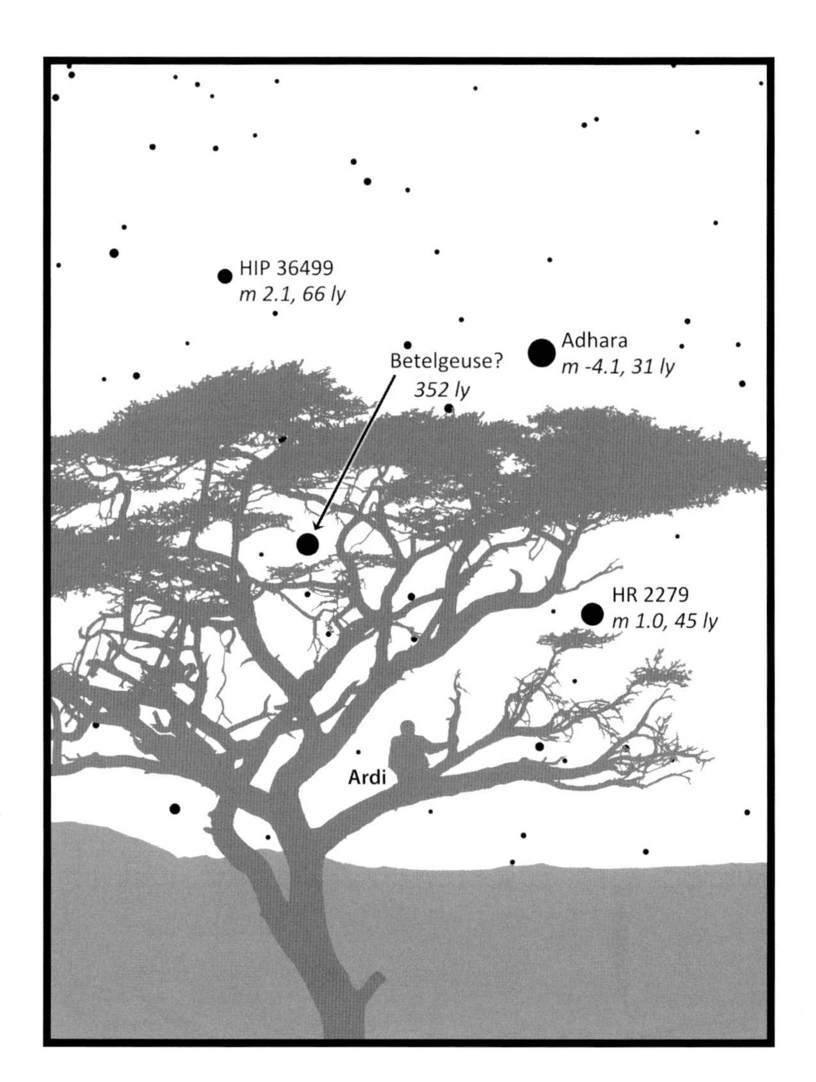

DIAMONDS IN THE SKY, WITH LUCY

3.2 Million Years Ago

Undoubtedly the most renowned paleontological find of an ancient hominid was the discovery of *Lucy*—a female *Australopithecus afarensis* who lived in East Africa more than three million years before our time. The news came as a bombshell in 1974. Her kind may have been direct ancestors of the genus *Homo,* and hence modern human beings. The history of human evolution was suddenly far more ancient than previously ever imagined. Lucy (whimsically named by her discoverers after the Beatles hit song "Lucy in the Sky with Diamonds") became a household name.

Lucy had a diminutive stature. At 3'8" tall, she was no larger than a chimpanzee. Males of her species were at least five feet tall. Her body was well-adapted for walking and running. What sort of lifestyle Lucy enjoyed in her environment is not entirely clear. Her teeth were small and unspecialized, suggesting that she pursued an omnivorous diet. But whether any meat that sustained her was secured through hunting, scavenging, or both activities, cannot be concluded with much certainty. We can be more assured that these small bipedal apes often themselves fell victim to predation.

In this scene we catch Lucy walking astride an ancient lake, bathed in the lunar light of the setting Moon as night gives way to dawn. What stars did Lucy see in her time? In the southwestern sky, a trove of bright stars glitters in the center of the portrait. These are the newly born stars of the *Orion Association*, which, in due time, will arrange themselves into the familiar hour-glass-shaped constellation of Orion, the Hunter. Rigel and Saiph, the feet of Orion, have already taken a firm stance parallel to the celestial equator. But Orion's belt—Alnitak, Alnilam, and Mintaka—is not yet properly strapped to his waist. Missing entirely are Betelgeuse and Bellatrix—the stars of Orion's shoulders. We encountered Betelgeuse in the previous scene, and if our projections hold true, it is still on its way to join Orion from a location just below the horizon. Bellatrix too is on its way over, coming in from the opposite direction.

Another pair of bright stars can be found in the upper left corner of the portrait: Mirzam and Nu Puppis. We alluded to Mirzam in the previous scene—it is the twin of Adhara. Mirzam has receded 100 light years in the previous million years, well on its way toward its modern position in Canis Major. Nu Puppis, another blue giant, will sink even further south to become a third magnitude star in the assembly of constellations that human sailors will one day fancy as a ship.

At Orion's feet is the distant supergiant Arneb, the brightest star in the constellation Lepus the Hare, in a location not far from where we find it in our own time.

The genus *Australopithecus* will continue to exist on Earth for another million years or so before giving way to hominid successors with more brainpower. These descendants will invent and make use of more sophisticated stone tools and weapons. They also will gain power over a valuable natural phenomenon with revolutionary consequences.

Prometheus Triumphant

1.5 Million Years Ago

Once upon a forgotten afternoon, deep in the hills of South Africa, a violent thunderstorm rolled over the land. As dark clouds gathered overhead, clamorous rumblings prompted animals to scurry for shelter. For the better part of an hour, flashes of intense lightning cut through the air. After the storm had cleared away, a pillar of smoke sputtered skyward from a distant hilltop. One intrepid creature quickly emerged from hiding. He set upon the hill with a purposeful gate. At the summit, he arrived upon a scene of scorched trees and blazing grass. Approaching the blaze cautiously, he prodded the dancing flames with the shaft of his wooden spear. The flames licked the spearpoint until it too caught the contagion of flickering light. He was mesmerized by his acquisition as nightfall set in. He then set upon returning home to the cave of his tribe, bearing his lit torch and the happy prospects of warmth, cooked food, and festivities.

Our hero was likely a *Homo erectus,* a species of hominid that flourished between 1.8 and 1.3 million years ago. The ability for controlled use of fire can be inferred from the burnt bones of antelopes and other animals found in caves that date back to this time.

In the Greek myth of Prometheus, the prize of fire was a stolen gift. Once again, art seems to imitate history. It is doubtful that anyone knew how to *make* fire more than a million years ago, so it had to be captured opportunistically—perhaps in ways resembling the narrative above. The ability to use controlled fire must have served the early hominids well. *Homo erectus* brought this capability to a wide range of territories that they eventually colonized throughout Asia and Europe.

And what of the fires in the sky—the stars—which adorned the heavens of this time? In this portrait we are gazing west from a southern latitude. On our left we find a dense field of stars that makes up a rich section of the southern Milky Way, which includes stars from the constellations Carina and Vela. On our right, we see a smattering of bright blue stars that belong to a widespread population of young hot stars in the *Scorpio-Centaurus Association.* This subgroup, designated *Lower Centaurus-Crux,* contains many stars from those two namesake-constellations.

The origins of the stellar associations, such as the one seen here and the *Orion Association* we just saw with Lucy in the previous portrait, follow typical patterns of star formation. Stars are born in groups when gigantic molecular clouds of cold gas collide together. Molecular clouds are brought into collision either where their orbits cross in Galactic spiral arms or by the pressure of shockwaves generated by exploding supernovae. When enough mass is squeezed into a small enough volume, pockets of gas will collapse to much higher densities under the weight of their own gravity. This leads to the creation of *protostars* in which the infalling gases are heating up by compression. If a protostar accumulates enough mass, the increased gas temperature induced by gravitational pressure will reach a degree that makes nuclear fusion possible, and a new star will *ignite.* Otherwise, the protostar may end up as brown dwarf—a failed star like those that orbit Epsilon Indi (page 16).

Thus the Galaxy at large is subject to bouts of severe weather. A Galactic storm made the world possible, while mundane storms bestowed a gift that would someday make civilization possible.

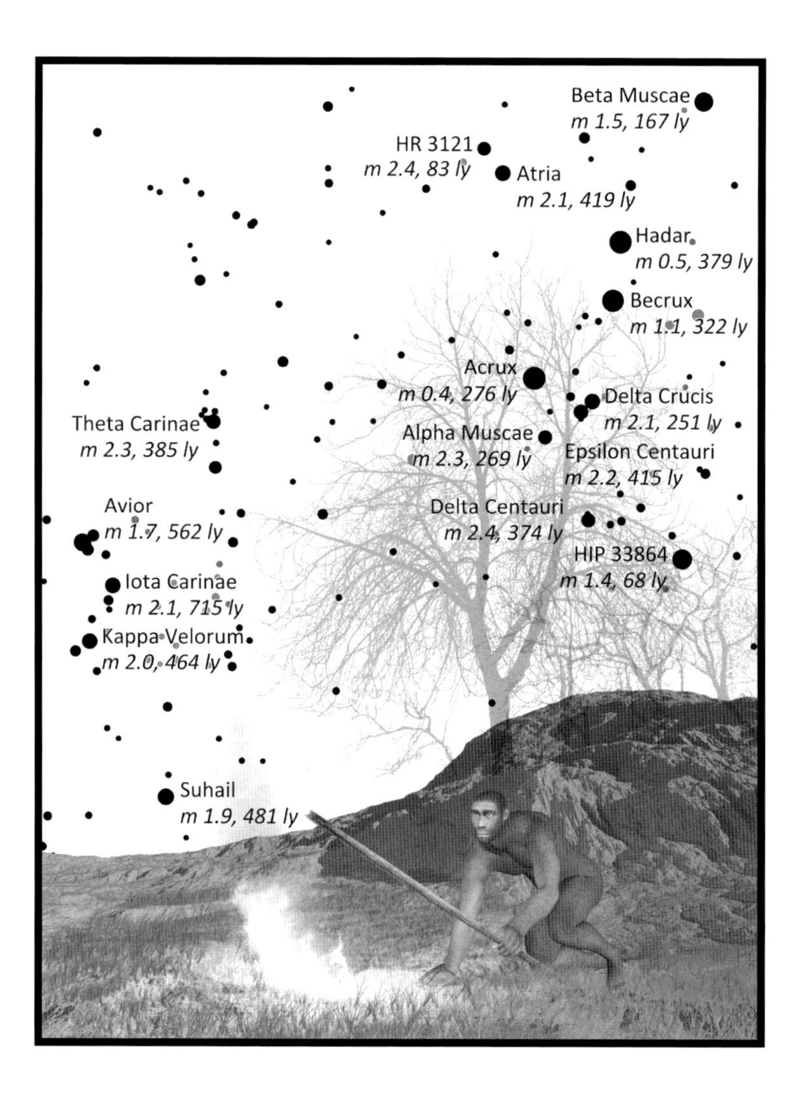

Hyperborean Safari

780,000 Years Ago

We now visit ancient Britain as it existed 780,000 years ago—a place very different from the island we know today. Lower sea levels, which then prevailed, left the isle conjoined with Europe by way of a broad land bridge that closed off the English Channel. It is possible that hominid pioneers traversed the bridge and began exploring southern England as early as 1.2 million years ago—contemporarily with their more clearly evident occupation of northern Spain.

By 800,000 to 700,000 years ago remains of stone tools were left behind in the region of Norfolk and Suffolk counties. A diverse range of exotic creatures populated these lands in this period: mammoths, rhinos, hippos, horses, hyenas, and saber-toothed cats that flourished in a Mediterranean-type climate. Early human pioneers would have had ample opportunities to conduct adventurous safaris at these northernmost borderlands of their wanderings.

A most curious geological event also occurred in this period. About 780,000 years ago is when the poles of the Earth's magnetic field last "flipped over." Such events, termed *geomagnetic reversals,* are rare, but they have occurred at least two dozen times in the past 5.3 million years. If we brought a magnetic compass back in time with us, we would find that its compass needle would point north in some epochs (as it does in ours) and sometimes south (as it would for many millennia before 780,000 years ago). Geomagnetic reversals do not happen instantaneously; they transpire over periods of hundreds or thousands of years. More commonly, a less drastic process occurs—a *geomagnetic excursion*—in which the strength of the magnetic field weakens and the directions of the poles wander for some time, but ultimately realign without reversing. Just how often such excursions happen is not well-known and impossible to predict. An imminent recurrence in the age of modern human civilization remains a distinct possibility.

In the scene presented here, we witness a stalwart pioneer standing on a cliff overlooking the ancient coastline of the North Sea. The ethereal multi-colored glow of an *Aurora Borealis* (the "Northern Lights") splashes across the night sky. Such a spectacle happens only rarely in modern Britain. But on the night depicted, with the wandering magnetic north pole just beginning to reassert itself, these apparitions could have been more frequent whenever the pole position swept near.

Turning our attention to the stars above, we see the Hyades star cluster setting in the west. The cluster has been on the outbound portion of its Galactic orbit for many millions of years, and its nearest approach to Earth occurred just 40,000 years beforehand (820,000 years ago). The Hyades cluster appears in this portrait nearly twice as close to us than it does in our time (80 light years versus 150) evidenced by its increased apparent size and brightness. The Pleiades star cluster is also visible in the upper left corner of the portrait. In our time, both the Hyades and the Pleiades are conspicuous parts of the constellation Taurus, but their positions in this scene correspond to the locations of Andromeda and Perseus respectively. We will observe these notable clusters again before our time-travel itinerary is over.

The pioneering efforts of Britain's first human inhabitants were brought to an end with the onslaught of an ice age. Subsequent ice ages would expel subsequent waves of immigrants. Contemporary Hyperboreans would hail from a new line of human descent.

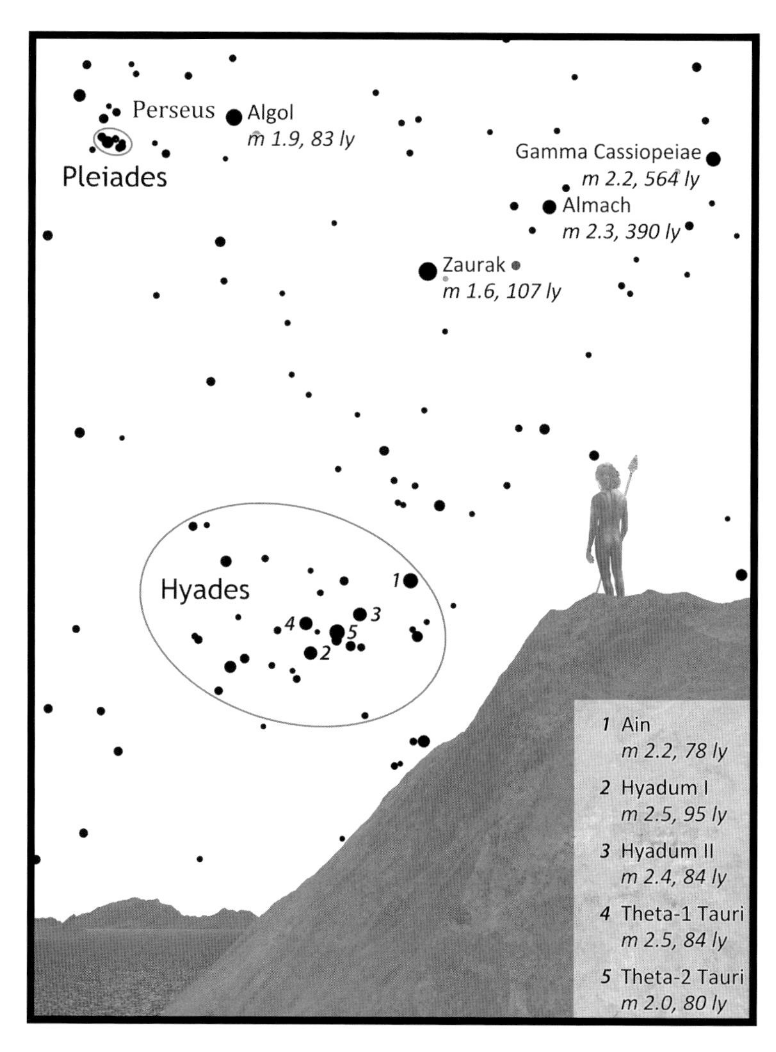

Mitochondrial Eve, The Greatest Grandmother of Humankind

206,000 Years Ago

So far in our time travels, we have encountered four different hominid species living in four different eras. Now we encounter the definitive ancestors of ourselves.

Homo sapiens, the species to which modern human beings belong, originated somewhere in East Africa. The earliest known fossil remains of anatomically modern humans date back to 196,000 years before our time. The defining characteristic of the ultimate human race is our large and highly developed brains, which facilitate complex language and thought. These abilities would eventually make it possible for us to flourish in modern civilization, but they were not so clearly advantageous in the period of our humble beginnings. Brainpower imposes a cost. The human brain consumes more metabolic energy than any other organ in the human body. So no matter how much more communicative, creative, and clever our ancestors were than their other hominid contemporaries, they also got hungrier. As they busied themselves with challenges of feeding themselves, they bore children infrequently. At that time, the total population of our *Homo Sapiens* ancestors may have been limited to just a few thousand individuals for many millennia—a far cry from the several billion people populating the planet in our own time.

Geneticists can claim with confidence that, among the individuals who founded the human race, there was one particular woman from whom all living human beings descend. The evidence is in our mitochondrial DNA—an isolated component of our genetic endowment that is inherited by each individual exclusively through one's mother. Hence, if you could trace the maternal lineage of your family tree (through your mother, your maternal grandmother, great-grandmother and so on) through some 10,000 generations, you would eventually discover a unique individual converged upon by every other person living today: *Mitochondrial Eve*—a woman who walked the Earth approximately 200,000 years ago.

In the scene depicted here, we find Mitochondrial Eve and her first daughter standing on a hilltop overlooking the banks of the Omo River in southern Ethiopia. In Eve's sky, we can trace the five main stars of the constellation Cassiopeia the Queen. They are strewn across the top and right sides of the portrait (see dashed line in chart). The nearest of those stars, Caph and Ruchbah, will join the other three to create the compact "W" asterism familiar to modern stargazers.

Arcturus can be spotted toward the upper left. The orange giant is one of the brightest stars in our time. But in Eve's time it is six times fainter and two-and-a-half times farther away.

Polaris lies very low in the northern sky at this sub-Saharan latitude. It is closely neighbored by the more conspicuous yellow giant star Capella, which appears slightly brighter and is 13 light years closer to Eve than it is for us today. Capella would have been the steadiest North Star for several generations of this era. Would it have inspired curiosity in the very first *Homo Sapiens?*

Eve's daughter points toward Capella's location. Northward lies the destiny for so many of Eve's descendants—out of Africa and into the world at large.

THE LAST NEANDERTHAL

37,000 Years Ago

We turn now to southern Spain. It is 37,000 years before our time. The world has been gripped by an ice age (or more scientifically speaking: a *glacial period*) for the past 100,000 years. Much of northern Europe has been buried in thick sheets of ice during all this time. Even the habitable ice-free areas remained much colder than they are today. Those who occupied them were a hardy lot—the Neanderthals.

The Neanderthals were an ancient race of people with stocky bodies that were well-suited for surviving cold climates. Evidence of their settlements in Europe date back to 300,000 years ago; their first arrival on the continent may have occurred much earlier. But by 40,000 years ago they faced competition from *Homo sapiens*—the latest out-of-Africa emigrants to Europe. The details of this confrontation are lost in the darkness of prehistory. But we certainly know that there were winners and losers. The Neanderthals lost.

The last Neanderthals in Europe died out 37,000 years ago. The scene depicted here portrays a lone survivor, hiding in the wilderness. Having kindled a campfire to remove the chill of the moonlit evening, he is pained with memories of the loss of his tribe. Only his canine companion, a lone wolf, gives him some comfort of camaraderie—and perhaps vice versa.

Beyond the icy mountains in the distance, we recognize the starry figure of Orion setting in the southwest. He too is hiding. In our time, the mythological hunter appears high above the southern horizon throughout most northern hemisphere locations. But in the time of the last Neanderthal, Orion stooped low through Spain.

This change in the constellation's position is not due to the motions of stars through space that we've grown to appreciate in our time-travel adventure. Rather, it is due to a wobbling motion of the Earth itself, *precession*. This wobble, induced by the gravitational forces of the Moon and Sun, gradually shifts the orientation of the Earth's axis in cycles of 25,772 years. Hence, the last Neanderthal lived nearly one and half cycles ago, when the north pole of the Earth's axis pointed toward an entirely different heavenly location—to a sky position near the bright star Vega.

The entire vault of the heavens reflected this axial shift, and so did the seasons. Summer began in December in the northern hemisphere; winter began in June (as they do in the Earth's southern hemisphere in our time). The location of Orion in the southwestern sky so soon after sunset tells us that it is early March. Autumn is imminent.

Orion too has a canine companion. Sirius, the "Dog Star," follows him across the heavens. In our time we would find Sirius below Orion's feet, but here we find it at the level of Orion's belt—this particular movement is due to the motion of the star itself. Procyon, a star whose Greek name means "before the dog," is also somewhat north of where it is sighted by modern stargazers; it can be spotted at the upper-left.

It is time now for us to move on. But before we leave the last Neanderthal behind, let us note that his story may not end precisely here. Recent analysis of DNA extracted from Neanderthal bones have revealed genes in common with modern Europeans and Asians—but not modern Africans. This suggests that Neanderthals interbred with *Home sapiens*. If so, it may be unfair to say that his race is entirely extinct. For if the roots of your ethnicity can be traced beyond Africa, a little bit of him may be living on through you.

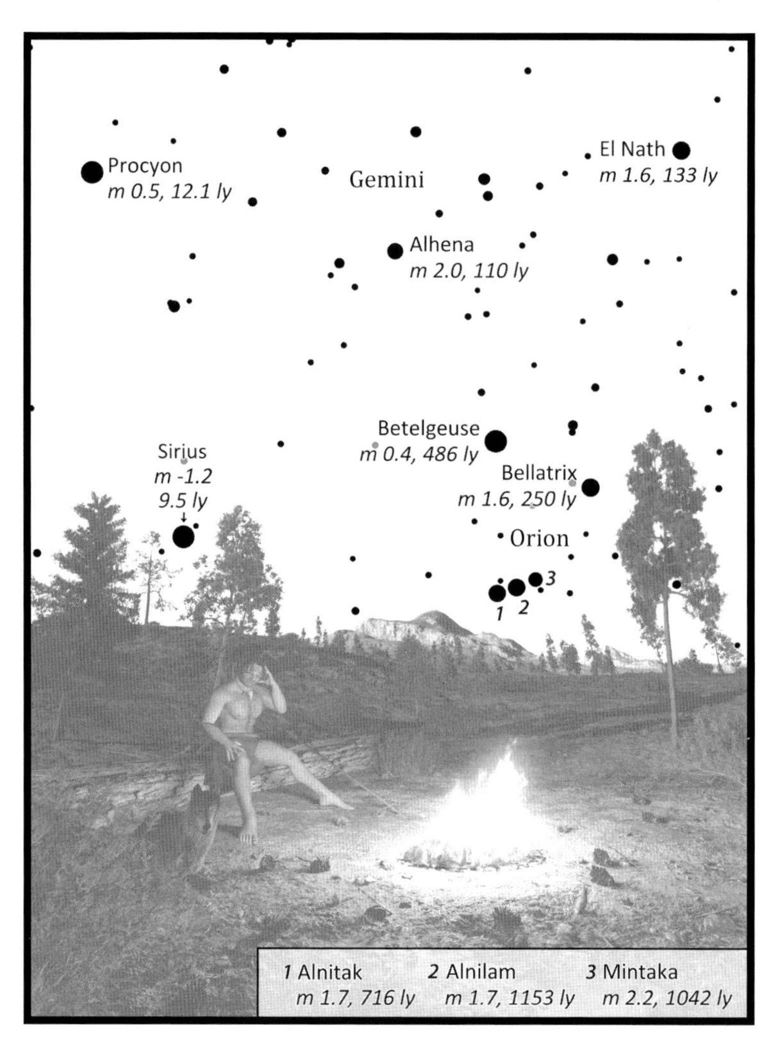

The Dawn of Civilization

4,500 Years Ago

We return once again to Britain. It is 2500 BCE. Gone is the ice age. An *interglacial* period has already begun. Europe thawed out 7,000 years beforehand. Modern humans are the only remaining people populating the Earth. More and more, they have been giving up their nomadic ways to become farmers and ranchers. They have settled into villages, and some villages have grown into cities. It is the dawn of civilization. The darkness of prehistoric times are giving way to a bright morning of recorded history.

Our knowledge of events taking place in the south of England, however, remain obscure. Written records are still lacking in this part of the world. But we are called to the region by one of the most celebrated archeological sites in existence: Stonehenge.

The origins of this stone monument date back to about 3100 BCE, when archeologists reckon that a circular embankment (a *henge*) was dug around the site, flanked on the interior by 56 regularly spaced holes. Various additions and modifications to the site were made in subsequent centuries, culminating circa 2500 BCE with the erection of enormous standing stones (some as tall as 22 feet and weighing as much as 45 tons) that are so iconic to Stonehenge. This grandiose feat of bronze-age engineering happened to coincide with another: the construction of the Great Pyramids in Egypt.

Which of the ancient peoples living in those days could have built Stonehenge and to what purpose cannot be well discerned. It may have primarily served some long-forgotten religious or ritualistic functions, as did so many other ancient architectural endeavors. However, Stonehenge incorporates certain design elements that also make it the earliest known astronomical observatory. The alignment of the stones allows observers to site the position of sunrise on the first day of summer—or, by looking through the opposite direction, where the Sun sets on the first day of winter. The site could have likewise helped a dedicated observer track the movements of the Moon and perhaps even predict eclipses.

In this scene we find Stonehenge as it may have looked when it was freshly built—4,500 years before our time. It is just before daybreak and we are facing due north. To our left, one of the five *megaliths* towers above us. The megaliths are arranged in a horseshoe configuration with the open end facing northeast, directed toward a position on the horizon where the Sun rises at the summer solstice.

In the sky above Stonehenge, we find all the constellations appearing in the configurations familiar to modern stargazers. The asterism of the "Big Dipper" (or the "Plough," as it is more commonly recognized in Britain) in Ursa Major is plainly visible in the upper right area of the portrait. In the upper center, we find a fourth magnitude star named Thuban in the constellation Draco the Dragon. Thuban was essentially the "North Star" in 2500 BCE, rather than Polaris. Once again, this shift is due to the Earth's axial precession, which we noted in the previous scene.

At the dawn of human civilization, our relationship with the celestial realm was commemorated by Stonehenge. As civilization advanced in sophistication, so too advanced the growth in knowledge of the heavens—along with ever-increasing astonishment.

Backyard Astronomy

Here and Now

"Contemporary civilization, whatever its advantages and achievements, is characterized by many features which are, to put it very mildly, disquieting; to turn from this increasingly artificial and strangely alien world is to escape from unreality; to return to the timeless world of the mountains, the sea, the forest, and the stars is to return to sanity and truth."

- Robert Burnham, Jr., Astronomer, 1966.

For all of human history, and untold generations of prehistory, people have looked up to the stars in wonder and fascination. In many cultural traditions, the stars were seen as pictorial representations of animals found in nature, characters from mythological stories, and man-made objects. This universal custom of tracing out constellations in the night sky fostered notions that the stars were placed up there by a divine artist or artists. The cosmic mural was perceived as some sort of spherical firmament encasing the Earth, shielding us from the secrets of the universe beyond it.

This perception began to be questioned soon after the celebrated astronomer Copernicus made fashionable the idea that the Earth orbited the Sun rather than the other way around. Somehow, another revolutionary idea also caught wind: that the Sun was just a nearby star, and all the other stars were faraway suns, strewn haphazardly throughout the unfathomable reaches of endless space.

Not everyone was so convinced. Some detractors were even well-educated. Johannes Kepler, the famous German astronomer who laid down important groundwork for comprehending the motions of planets, was dismayed and disoriented by the idea of an unlimited universe. In 1606 he complained, "This very cogitation carries with it I don't know what secret, hidden horror; indeed one finds oneself wandering in this immensity having no limit, no center, and no determinate places."

Kepler's objections notwithstanding, the age-old notion of the solid "celestial sphere" crumbled irretrievably in 1609, when Galileo began recording observations with the astronomical telescopes he built.

Galileo's gifts to science ultimately brought astronomy, the oldest of sciences, within reach of everyone. No one ever forgets their first experience of seeing celestial objects through a telescope. To behold, as Galileo first did, the mountainous craters of the Moon, the banded disk of Jupiter attended by its neat row of satellites, the rings of Saturn, and the phases of Venus is a rite of passage that rarely fails to instill a vivid impression of the wonders of the solar system.

Such visual enticements often blossom into a lifelong hobby. For a modest investment, one can easily acquire an instrument even more powerful than Galileo's best spyglass. The splendors of the universe beyond the solar system—double stars, star clusters, nebulae, and galaxies—lay bare and open to exploration. One can undertake such excursions from any place on Earth with wide horizons and dark skies, whether it be some remote campsite in the wilderness or your own backyard.

The stories told about the stars in contemporary civilization are now very different from the tales of ancient times. The old star lore celebrated a mythical past. The new star lore, namely science fiction, offers visions of human destiny. Will we fulfill them?

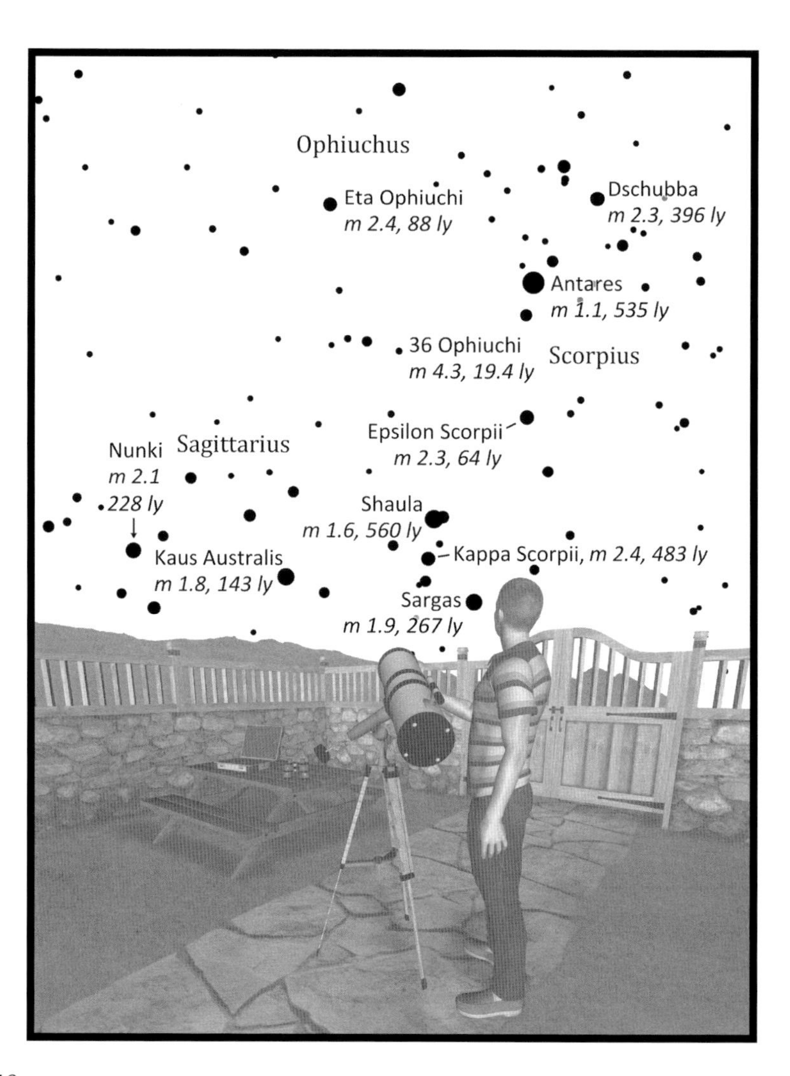

END OF THE INTERGLACIAL

50,000 Years Ahead

In the course of our 5.3-million-year-long journey, the Earth's climate has been decisively trending cooler. At the beginning of the Pleistocene epoch, 2.5 million years ago, global climate even became fickle. Extended periods of glaciation became the norm. Continental ice sheets repeatedly covered up North America and Europe, transforming their landscapes into vast arctic wastelands for tens of millennia at a stretch. These *ice ages* were punctuated by warm spells that compelled the ice sheets to retreat—the *interglacials*. But they were only temporary. Throughout the present ice age, interglacials have been in effect for only 10 to 15 percent of the time.

Human civilization has had the good fortune of growing up in what perhaps will turn out to be an usually long interglacial. It has already lasted more than 12,000 years and may continue on for another 50,000 years. This prediction is based on an astronomical theory: Milankovitch cycles, which link ice ages to long-term changes in the Earth's axial tilt and orbit.

We have recently noted axial precession—the wobbling of the Earth's rotational axis in cycles of 25,772 years. The Earth's axial tilt also varies. In our time, the axis is inclined 23½°. But when Stonehenge was built it was 24° and when the last Neanderthal died in Europe, it was closer to 23°. Overall, it varies between 22° and 24½°, in cycles of about 41,000 years.

Earth's orbital path around the Sun is also subject to change, due mainly to gravitational perturbations from Jupiter and Saturn. The Earth's orbital eccentricity ranges between 0.5% and 5.8% in 413,000-year cycles. In our time, the eccentricity of the Earth's orbit is only 1.7%, which describes a nearly perfect circle. When eccentricity is low, the distance between the Earth and Sun varies little from season to season—differing by three million miles in our time. At the maximum 5.8% eccentricity, however, the annual variation in the Earth-Sun distance stretches out to 11 million miles. Variations in orbital speed are also accentuated by high eccentricity, so that the lengths of the seasons become more uneven. When the Earth is nearest to the Sun (*perihelion*), it rushes by even more quickly and when it is farthest from the Sun (*aphelion*), it ambles along even more slowly.

Milutin Milankovitch was a Serbian engineer and geophysicist who, in the early 20th century, worked through these considerations to calculate how much solar radiation has been received by the Earth—particularly in the more sensitive land-dominated northern hemisphere—over geological timescales. It was a marvelous achievement for someone working before the days of electronic computers. But did his theory explain the ice ages? His results were disputed for many decades, even after his death, but Milankovitch was ultimately vindicated by the evidence of ocean-floor sediments. Core samples indicate climatic cycles in their strata, which were indeed synchronized with the Earth's orbital parameters.

We can also use Milankovitch's method to estimate when the next ice age is coming. A recent estimate puts that date at 50,000 years from now. How will our distant future descendants adapt to the next ice age?

In this frigid scene, we espy a futuristic observation post on the west coast of Canada. Will the inhabitants of a re-glaciating Earth seek refuge elsewhere? Alpha Centauri, our nearest celestial neighbor, has in all this time migrated out of the southern skies to the celestial equator, where it can be sighted from locations throughout the entire globe. It seems to beckon humanity to the stars.

The Death of Betelgeuse

500,000 Years Ahead

We previously observed the giant star Betelgeuse in Ardi's time (page 36) when it was a whiter star in an earlier stage of its life. We also observed it from ancient Spain (page 46), in its current red giant phase—near the end of its life. Sometime in the next one million years, we cannot know exactly when, Betelgeuse's starry existence will come to a spectacularly violent end as it erupts into a Type II supernova explosion.

Astronomers estimate that supernovae of all types occur, on average, about once every 50 years in our Galaxy. Some of the most recent ones that have been recorded in history include one in 1604 (recorded by Johannes Kepler), 1572 (recorded by Tycho Brahe), and 1054 (recorded by Chinese astronomers). The nebulous remnants of these explosions can still be observed in telescopes today. Since 1604, other supernovae that may have exploded in the plane of our Galaxy could have easily escaped our notice if they did so in other Galactic quadrants—far enough away to be obscured by intervening dust. In 1987, a supernova occurred in the Large Magellanic Cloud—a satellite galaxy that orbits the Milky Way—making it the closest supernova to Earth observed in recent times.

Betelgeuse's date with fate is depicted here occurring a half-million years after our time. When Betelgeuse goes supernova, it will outshine the combined light of all the other stars in the Galaxy. Seen at a mere distance of 500 light years, its apparent brightness on Earth will rival the brilliance of the full Moon.

Looking straight up in the sky, we observe the star we once knew as the shoulder of Orion (a somewhat dislocated shoulder by this time). Our vantage point is a conifer forest at high elevation and low latitude. The tops of the tallest pine trees are illuminated by the ruddy glow of evening twilight.

Elsewhere in the portrait, we find many familiar stars in unfamiliar locations. Vega and Altair, two of the three stars of the "Summer Triangle" asterism, are positioned in what northern hemisphere dwellers would recognize as the "winter" sky. Pollux and Castor, the twin stars of Gemini, are going their separate ways. El Nath, the tip of one of the horns of Taurus the Bull, is thrust into Orion's neck. The Hyades cluster has continued to recede away from the solar system, diminishing in both apparent size and brightness.

Betelgeuse's demise surely won't be the nearest supernova in Earth's history. The discovery of radioactive iron deposits in the crust of rocks on the deep-sea floor of the Pacific Ocean could be evidence of a previous supernova that exploded less than 100 light years away from Earth just a few million years ago. Given that the entire natural history of the Earth stretches back some 4.6 billion years, it is reasonable to suppose that even closer supernovae—close enough to affect or even endanger life on Earth—were recurring events. Certainly they will occur again, but probably not anytime soon. Once every few hundred million years is a reasonable guess.

These dangerous aspects of supernovae, however, tell only half the story. In another respect, they are life-bestowing. Supernova explosions create heavy elements and scatter them around the Galaxy—including elements that are necessary for biochemistry. Hence, past generations of dying stars made essential contributions to the chemical composition of our solar system's planets at the time of their formation.

Other stars may interact with the solar system in an even more direct fashion, as we shall consider next.

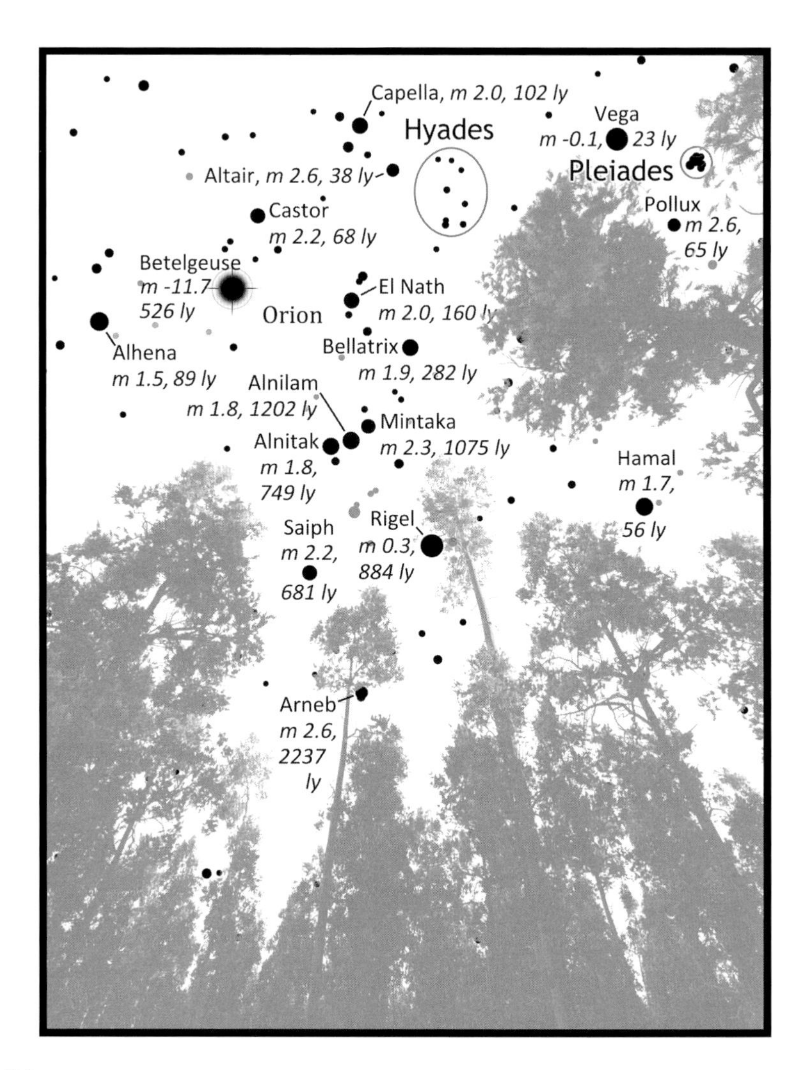

Encounter with Gliese 710

1.4 Million Years Ahead

While it is not possible to pin down when Betelgeuse might go supernova within a time frame narrower than a million years, other celestial happenings can be forecast with much better precision. We can be quite confident, for instance, that the Hyades star cluster passed within 80 light years of the Earth about 820,000 years ago. We may also know when other nearby stars will encounter the solar system in the future.

One such star, Gliese 710, seems to be an unremarkable orange dwarf star shining at magnitude 9.7 from 64 light years away in the constellation Serpens. Yet, it will approach the solar system more closely than any other star that we currently know of. In 1.4 million years, the star may pass within 50,000 AU—less than one light year distant.

No star is that close to us today (except the Sun, of course). In our time, the Alpha Centauri star system is 4.4 light years away (page 10). Gliese 710 will be five times closer than that, or just 1,650 times further than Neptune. This is still a considerable distance on a planetary scale, but it may be close enough to penetrate the Oort Cloud—a theoretical reservoir of cometary material which may surround the solar system out to a radius that extends perhaps as far as 100,000 AU away.

What Earthly consequences might transpire from such a close encounter? A star that penetrates the Oort cloud may potentially perturb its contents in a manner that diverts comets into the planetary region of the solar system. However, Gliese 710 is not thought to be massive enough to do so. On the other hand, Gliese 710 may itself have a cometary reservoir analogous to the solar system's Oort cloud. If so, then the Earth will have to pass through *it*.

When Gliese 710 arrives, it will shine conspicuously at magnitude 0.1 in the Earth's northern skies (conversely, the Sun will reach magnitude -3.3 in Gliese 710's skies).

In the scene depicted here, Gliese 710 can be spotted near the upper center of the portrait. Our view is directed northeast, early one summer morning off the coast of an Aleutian island in Alaska. A future species of Humpback whale playfully breaches high above the sea surface—perhaps glimpsing the celestial view as he towers momentarily skyward.

Returning our own attention skyward, a comet is seen streaming overhead. Its long tail, blown back by the flow of the solar wind, is directed away from the Sun. Though it seems to be standing still, it is moving through the solar system much more rapidly than a typical comet. Could this be an alien comet? An omen brought to us from the star system Gliese 710? Perhaps.

Comets, regardless of origin, could have once played a crucial role in the composition of planet Earth. It is widely believed that some proportion of the existing water in our oceans must have been seeded by icy comets that had repeatedly struck our planet during the earliest epochs of the Earth's formation. Since the oceans made it possible for life to flourish on Earth, we may add *comets* below *supernovae* on our growing list of heavenly phenomena to which we owe our very existence (a theme that we shall explore further in our fourth itinerary).

The solar system will rendezvous with many other Galactic stars in future eons. Let us now leap far ahead to witness another close encounter.

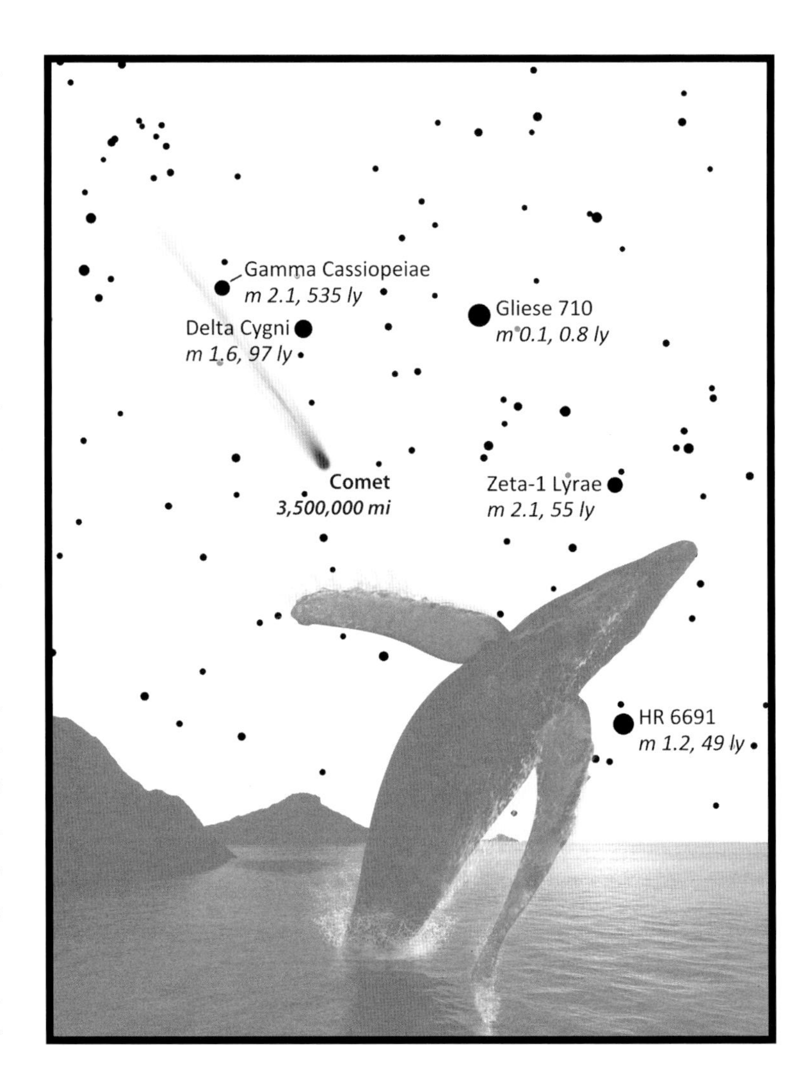

The Moon and Albireo Over New Hawaii

5.3 Million Years Ahead

If you ever meet up with an amateur astronomer manning a small telescope on a summer evening (like we did on page 50), and you ask to be shown an example of a double star, chances are that your attention will be directed to Albireo. The star's contemporary position in the constellation Cygnus makes it "the eye of the swan." But with a modest amount of magnification, it resolves into *two* eyes—a bright one, Albireo A, which is amber in color, and a fainter one, Albireo B, which is blue.

The pair is widely separated in physical space—by at least 4500 AU. But they are still loosely bound to one another by their mutual gravity. The orbital period of the components around their common center of gravity is too slow to be precisely estimated, but it likely exceeds 75,000 years.

The system's orbit around the Galaxy at large is somewhat less eccentric that that of the Sun's. Albireo comes as close as 20,600 light years to the center of the Galaxy at minimum, and as far as 24,000 light years at maximum, where we find it today. This is where its orbital speed is slowest, while the Sun's current position near the innermost part of its orbit is close to where the Sun goes fastest. Hence, the Sun is overtaking Albireo rapidly. The 430 light years separating the two systems in our time will close to less than 25 light years in 5.3 million years.

We witness the Sun's encounter with Albireo in the scene depicted here. In the intervening time, Albireo's position has migrated south to a zodiacal region previously occupied by Antares. The brightness of the component stars have increased 300-fold. The angular separation of the pair has widened into a *naked-eye double,* spanning a third of the width of the Moon.

As with our visit to Gliese 876 (page 20) we opt for a telephoto view. Once again, this spreads the background stars thin, but it nicely exhibits Albireo's conjunction with the waxing crescent Moon.

In the foreground looms a small volcano, rising above the sea. Its location is at the easternmost end of Hawaii, and its eruptions mark the creation of a new island that did not yet exist in our time. The rest of the Hawaiian Islands would be unrecognizable to us. The summits of Haleakala, Mauna Loa, and Mauna Kea have been reduced to three small isles far to the west. All the islands west of Maui that we knew of in our time have long since sunk below the sea.

The rapid geographic flux of the Hawaiian islands is due to a *volcanic hotspot* (which creates new islands), plate tectonics (which scoots them westward at a rate of a few inches per year), and severe weathering (which washes them back into the sea). We may also note that the big island in the sky, the Moon, is also migrating. It has been climbing into a higher orbit at the rate of 1½ inches per year (125 miles in 5.3 million years) by robbing energy from the Earth's rotation (a day on Earth will be two minutes longer by then).

We have now followed the course of Spaceship Earth for 10.6 million years. In that time, the solar system has traversed 8,000 light years of Galactic space. Let us now take a detour through Earth's distant future across an even larger leap in time.

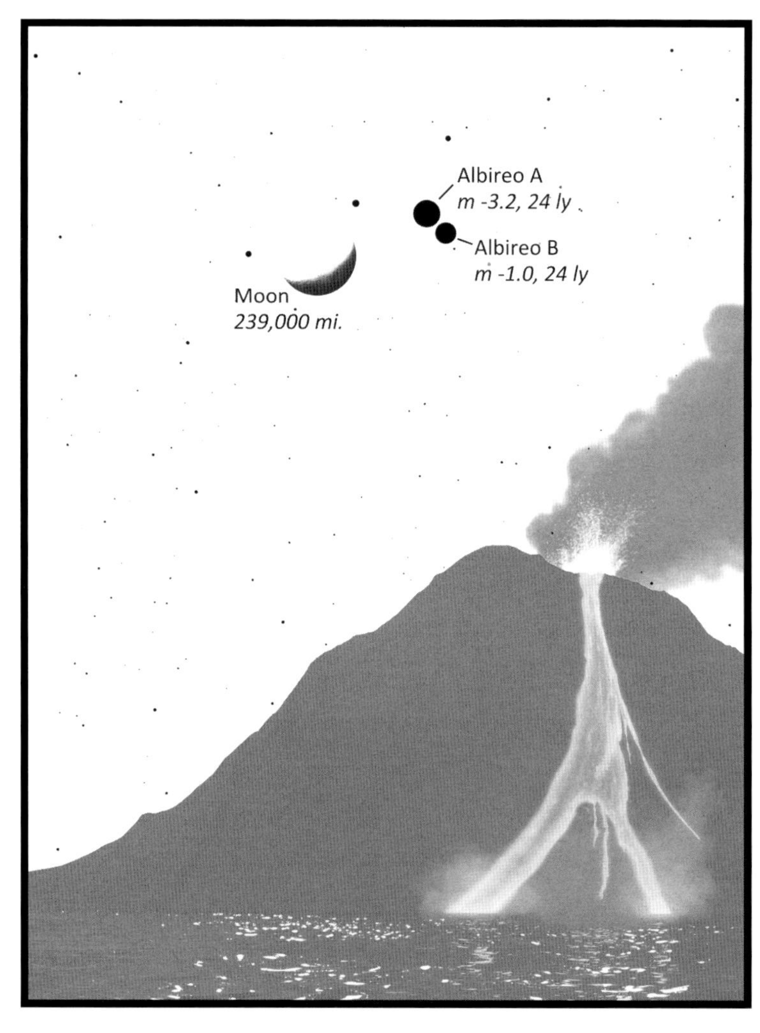

Detour: The Chilling Stars

25 Million Years Ahead

In the ten million years we spent aboard Spaceship Earth, the solar system has traversed only a small percentage of its Galactic orbit, which takes 300 million years to complete. Roughly 25 million years from now, our Sun will reach its closest proximity to the center of the Milky Way. From there it will start migrating out toward the edge of the Galaxy while its orbital velocity begins to slow down. As it does so, the Sun will exit the Orion Spiral Arm, where it had been spiralling inward for the previous 150 million years. On its way out, the Sun must pass through the arm's outer rim. This is a region densely packed with newly born massive stars; it contrasts sharply with the arm's inner rim, which is clogged with the dark nebulae of condensing gas clouds (see page 40). A photograph of the Whirlpool Galaxy (right) depicts the contrast nicely—as do most photographs of other two-arm spiral galaxies oriented face-on.

The Sun will also plunge through the Galactic mid-plane roughly 25 million years from now—a natural consequence of the Sun's oscillating motion perpendicular to the Galactic plane (illustrated below). The Galactic mid-plane is also a region of maximum star density. The last time the Sun crossed through it was when we saw diamonds in the sky with Lucy (page 38). Next time, the diamonds will be even more abundant.

This is the scenario depicted in the portrait. An accompanying finder chart is not provided this time; it would be of little use to us. None of the visible stars have proper names. It will suffice to know that the four bright stars on the lower left belong to the star cluster Roslund 5, which the Sun will intercept in this epoch. The four bright stars to the center and right are cataloged field stars that the Sun is overtaking. And the trio of brightest stars on the upper right are young stars that do not even exist yet in our time—a new belt for a future Orion.

We take in the scene from an aerial vantage point that places us near a *stratovolcano* at the tip of a Pacific island west of Papua New Guinea. On this moonlit night, the sea is clouded over and the protruding volcano is covered with snow and glaciers. This is a perilous juncture in the ongoing journey of Spaceship Earth. Supernovae are common on the outer rim, flooding the solar system with cosmic rays. When these high-energy particles reach the Earth's lower atmosphere, they catalyze cloud-seeding. The increased cloud cover boosts the reflectivity (or *albedo*) of our planet, so that more incoming sunlight is deflected back into space. Shielded from the Sun, a global chill has set in—perhaps more severe than any ice age that has occurred for hundreds of millions of years. But even this is not the end of the world. The clouds will eventually dissipate. Shrugging off the chill, the Earth will carry on for eons that we cannot even begin to fathom.

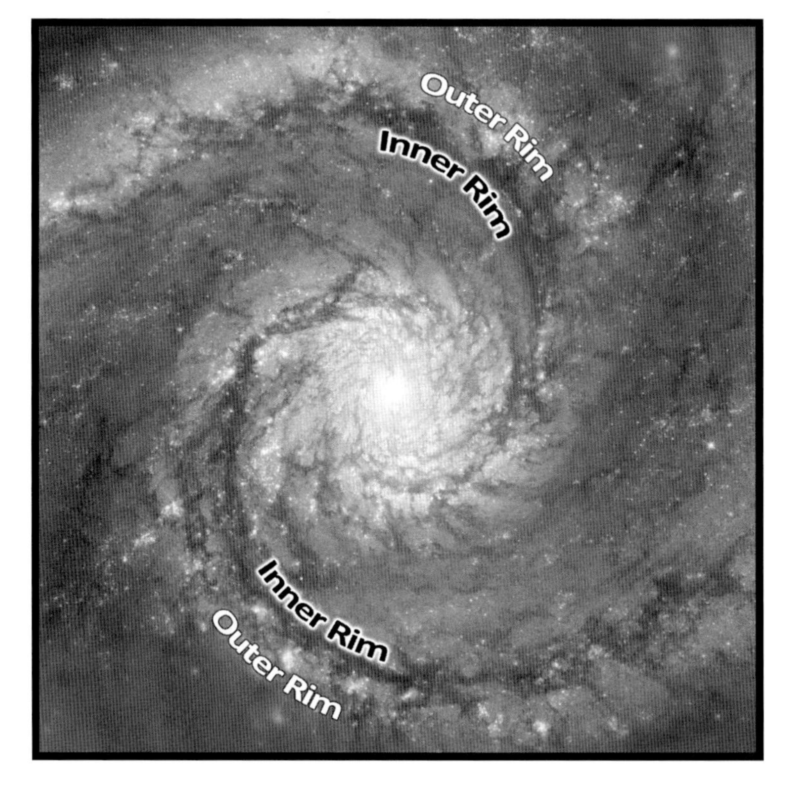

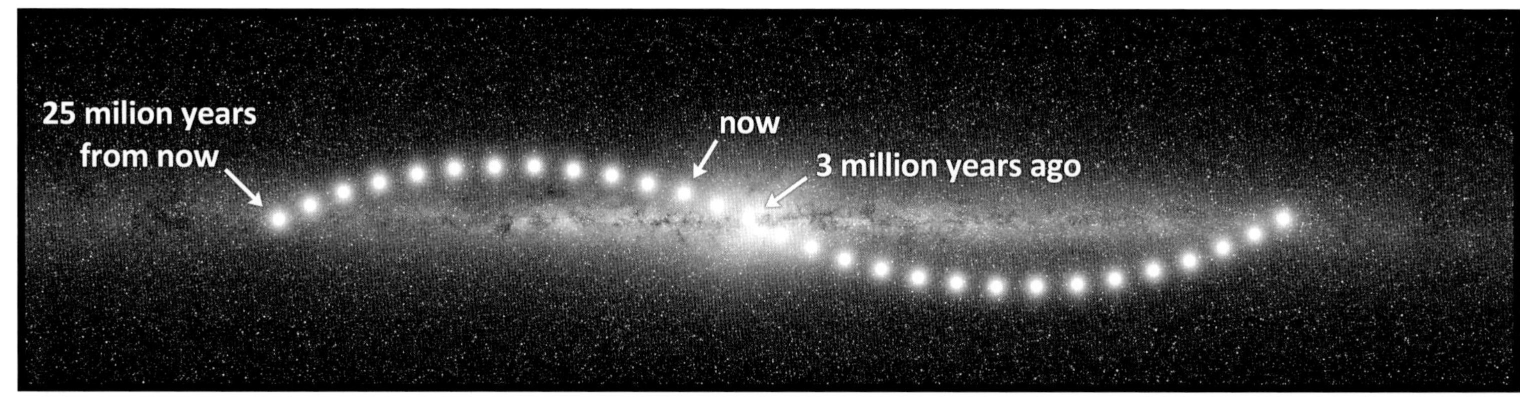

THIRD ITINERARY

The Quest for Planets Beyond the Solar System

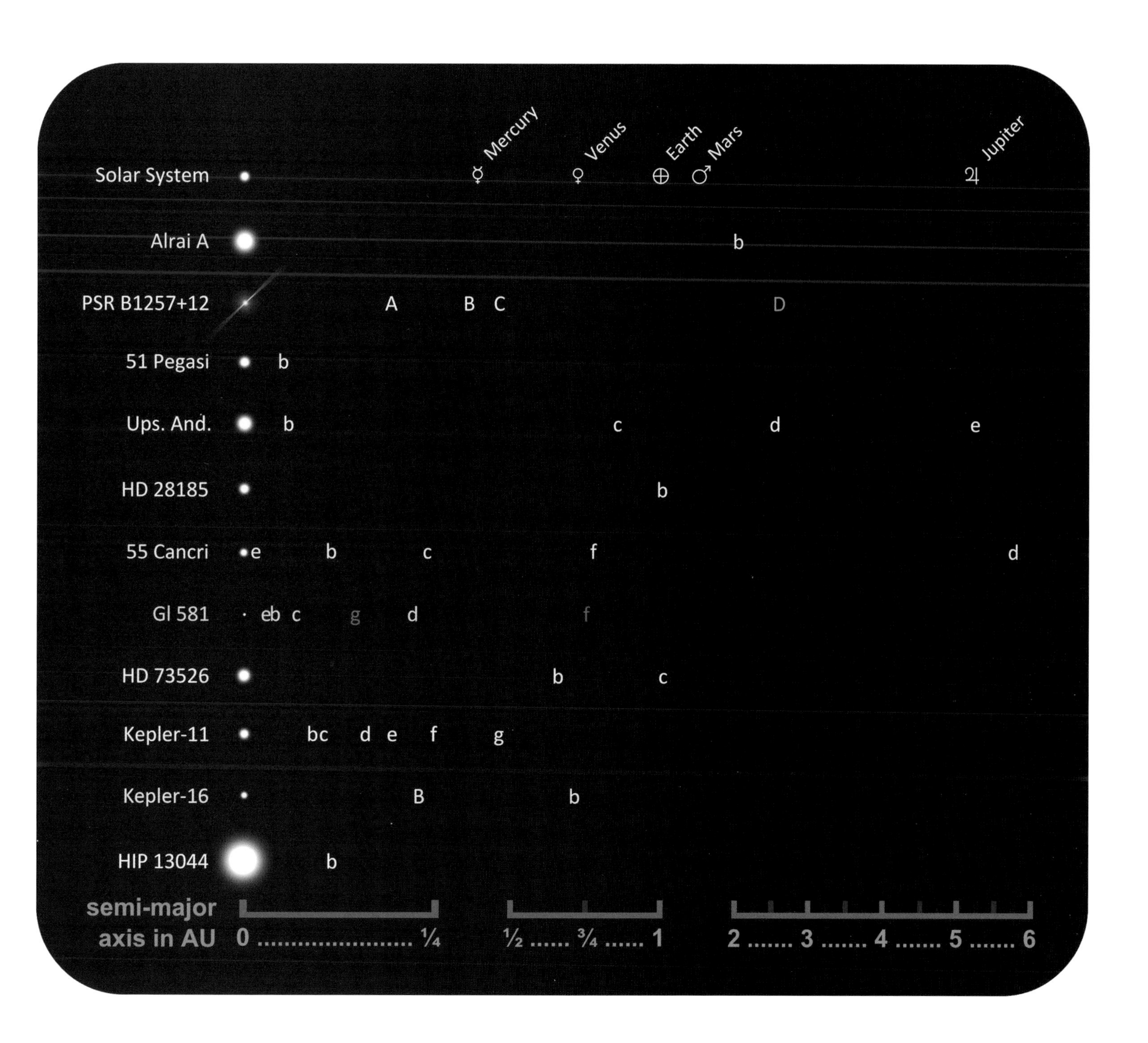

EXOPLANETS

Old Concept, New Science

The discovery of worlds beyond the solar system (termed *extrasolar planets,* or *exoplanets* for short) is an exciting recent development in modern astronomy. However, the notion of other worlds beyond our own is not a new idea. The Greek philosopher Democritus, who flourished in the 5th century BCE, supposed that an infinite number of worlds must exist. Moreover, he asserted these worlds were assimilated and dissimilated by the motions of atomic elements—a very modern idea. Democritus' opinions were rejected by Plato and Aristotle, but later reprised in the Hellenistic era (4th to 1st centuries BCE) by Epicurus and Lucretius. These latter authors inspired Giordano Bruno in the 16th century CE to herald a similar cosmology. Branded as heretical by religious authorities, Bruno's outspokenness ultimately cost him his life.

However, with the advent of the telescopic era of astronomy in the wake of the Copernican revolution, it was already becoming clear in the 17th century CE that the planets known to the ancient Greeks were not merely "wandering stars," but worlds unto themselves. The known solar system grew even larger with the discovery of additional planets, Uranus and Neptune, in the 18th and 19th centuries. Serious attempts to find planets outside the solar system were also made in these centuries. Perceived anomalies in the orbit of the binary star 70 Ophiuchi, for instance, led some astronomers to propose an unseen perturbing planet, but it proved impossible to confirm any of these claims—not then nor even now.

In the 20th century, astronomers securitized the positions of very nearby low-mass stars, recorded on series of photographic plates, for the signature of any side-to-side "wobbles" induced by gravitational tugs of orbiting planets. Claims of detected planets by this *astrometric* method for Barnard's Star, Lalande 21185, and 61 Cygni were asserted in this period, but these attempts too did not stand up to the test of time.

Later in the 20th century other astronomers sought to detect the same sort of wobbling motion but in another dimension—to and from the observed line of sight. These *radial velocities* can be inferred from Doppler shifts in spectrographs.

Spectrographic precision improved markedly in the age of electronic sensors, bringing minute velocity changes within range of detection. Another advantage of spectrographic detection of radial velocities is that its precision is not diminished by star distance; it is mainly limited (for a telescope of a given aperture) by a star's apparent brightness.

A disadvantage of radial velocity measurements, however, is that the motions they capture are one-dimensional. Consequently, the prospective planet's *orbital inclination* (the "tilt" of the orbit relative to the observed line of sight) is undetermined. Without this information, one may only set a lower bound to a planet's estimated mass, based upon the extreme case that the planet's orbital plane is aligned to us exactly "edge-on." It is only possible to fix an upper limit to a planet's mass estimate by tracing a star's wobbling motions in at least two dimensions.

We have already visited one nearby star with known exoplanets on our first itinerary—Gliese 876 (page 20). On this next voyage we will encounter several more and get a broader sense of what other star systems are like. Many surprises are in store.

ALRAI Ab

First Detection of a Confirmed Exoplanet

Between 1980 and 1987 sixteen bright stars were put under continual surveillance by a team of astronomers based in Canada. Their high-resolution spectrograph was capable of detecting radial velocity changes as slow as 12 meters per second (28 miles per hour). The brightest star in their observing program was Alrai A (also known as Gamma Cephei), an orange subgiant star 46 light years away in the constellation Cepheus the King.

A conspicuous change in the star's velocity was recorded, induced by its known companion star, the red dwarf star Alrai B. However, after subtracting this motion out of the data, another, smaller motion was left over—a variation of about 25 meters per second (56 miles per hour) occurring in cycles of approximately 2½ years.

In 1988 the team reported that a planetary body in the Alrai system was probable. But over the next few years another possibility was favored—that Alrai A was pulsating, and the changes in radial velocity detected were merely due to the star itself growing and shrinking in size. After another ten years slipped by, more spectrographs of the star were made at higher precision. Once these new measurements were reported in 2003, the pulsation scenario was ruled out and a planet was ruled in. The planet is Alrai Ab—a gas giant at least twice the mass of Jupiter, orbiting the star at a distance of 2 AU every 903 days.

Alrai Ab, then, was the first confirmed exoplanet ever detected, even though it was not the first exoplanet ever confirmed! That latter honor was bestowed quite unexpectedly in 1992 to a pair of faraway planets that we shall explore next. But the chronicles of Alrai Ab arouse a knotty question: when, exactly, can one say that a planet has been *discovered?*

Careful readings of history reveal that groundbreaking scientific discoveries are not typically neatly defined events. More often they are protracted processes of delayed recognition. Uranus, for example, had been observed many times over the span of nearly a century by several eminent astronomers before being recognized as a planet. John Flamsteed perhaps recorded its existence first when, in 1690, he mistook the object as a star and bestowed it with the designation "34 Tauri" in his star catalog. When William Herschel stumbled upon it in March of 1781, using a telescope of his own design that was finer than any instrument previously constructed, Uranus was discerned to be a disk. Upon observing it again on subsequent occasions, its motion against the more distant background stars also became apparent, prompting Herschel to announce the discovery of ... a new comet! Fitting the object to a cometary orbit, however, proved to be impossible. But when the mathematical astronomer Anders Lexell combined Herschel's observations with a previously unrecognized observation of the object, made in 1759, he came up with a solution for a planetary orbit. By 1783, Lexell's solution was universally accepted and the newly discovered object, ultimately designated *Uranus,* was duly recognized as the planet we know so well today.

The story of Neptune is a saga that spans even more time. Galileo surely observed it in 1613 when he sketched a background star at its position near where Jupiter happened to be—234 years before the date of Neptune's "official" discovery. It makes one wonder what secrets of the universe are hiding in plain sight even today.

On the other hand, it also happens on rare occasions that an unexpected finding leaps out at us.

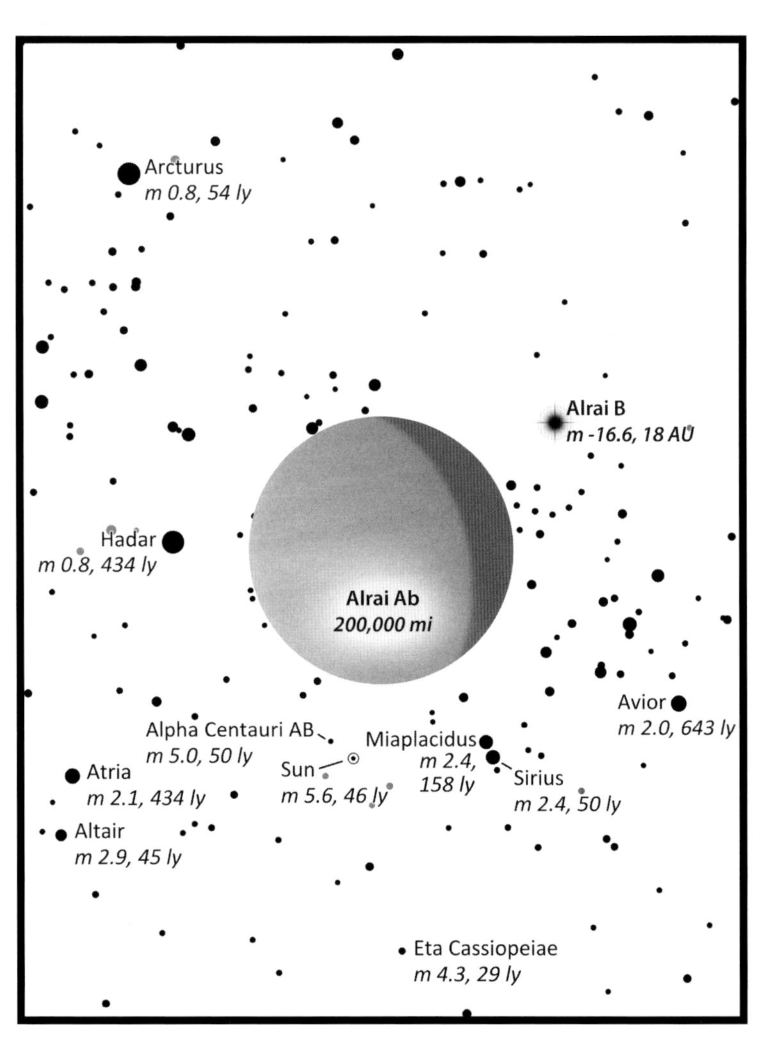

PSR B1257+12 B AND C

First Confirmed Exoplanets

The first exoplanets to be recognized as such are two exotic worlds orbiting a pulsar situated high above the plane of the Galaxy. The path to their discovery, unlike the example of Alrai Ab, was spectacularly rapid.

Pulsars are a variety of neutron stars—the small core remnants of supernova explosions. Neutron stars are composed almost entirely of densely packed neutrons—hence the name. They are analogous to white dwarfs (see page 12) but are even more extreme. A typical neutron star packs up to two solar masses into a sphere as small as 15 miles in diameter. Whereas a teaspoonful of white dwarf material would weigh fifteen tons on Earth, a teaspoonful of neutron star would weigh *four billion* tons.

Neutron stars may also rotate extremely rapidly—up to several hundred rotations per second. Pulsars are highly magnetized neutron stars which emit corotating beams of radiation in the manner of a lighthouse beacon. If the plane of rotation happens to be aligned with the Earth, the beams can be detected as rapid pulses by radio telescopes.

The pulsar designated PSR B1257+12 was discovered with the Arecibo radio telescope (the largest dish antenna in the world) in 1990. The object was found in the zodiacal constellation Virgo the Virgin. The distance of the pulsar from Earth eludes certain measurement. It is roughly estimated to lie 1,000 light years away.

PSR B1257+12 spins at a rate of 160 rotations per second. Pulsars, by their nature, are highly reliable clocks, so it was very curious that the timing of PSR B1257+12's pulses were found to be slightly irregular. But subtle changes in its pulsation rate may, in accordance with the Doppler principle, correspond to changes in the object's radial velocity. Hence, it was soon surmised that PSR B1257+12 wobbles to and fro. In 1992 the pulsar's discoverer and a colleague reported that the wobbling motions were consistent with gravitational tugs of two orbiting bodies, which history records as the first exoplanets ever discovered.

These planets, PSR B1257+12 B and PSR B1257+12 C, are similar to each other in size (4.3 and 3.9 times the mass of Earth respectively). They were likely formed from the wreckage of the destructive supernova explosion that created the pulsar—reincarnations of previously existing planets of the previously existing star. A smaller third planet, A, at least twice as massive as the Earth's Moon, was subsequently discovered in 1994, and a fourth body, D, having the mass of a dwarf planet or large asteroid was provisionally detected in 2002.

In the portrait depicted here, we view planet C from up close. Iridescent ribbons of an aurora encircle its magnetic polar region. The abundance of charged particles emitted from the energetic pulsar would create auroras continuously on any planet in the system that has a magnetic field.

The stars we see arrayed in the background are as they would appear from our vantage point far removed from the Galactic plane. Our distant Sun, too faint to be seen at its apparent magnitude of 12.2, is lost amongst the more conspicuous apparitions of some of the most luminous stars in the entire Galaxy. These include Rigel and Betelgeuse in Orion (at top), Canopus and Spica (center), and Antares and Shaula in Scorpius (at bottom).

Most confirmed exoplanets discovered to date, however, are located much closer to home.

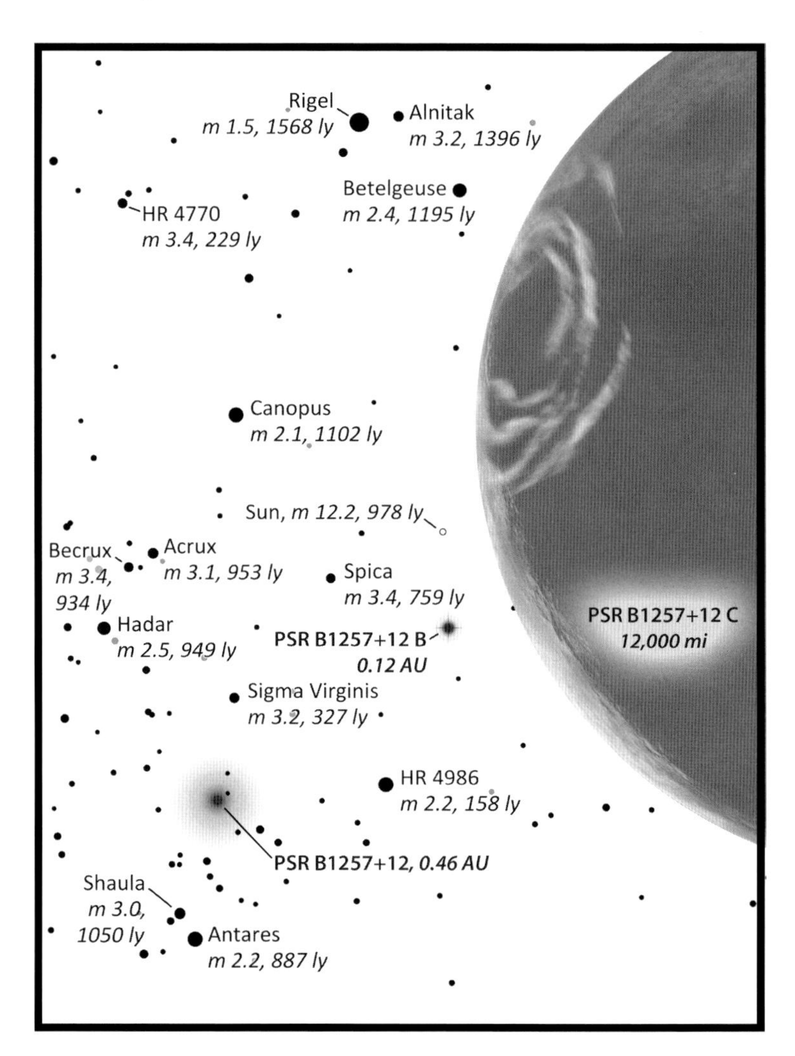

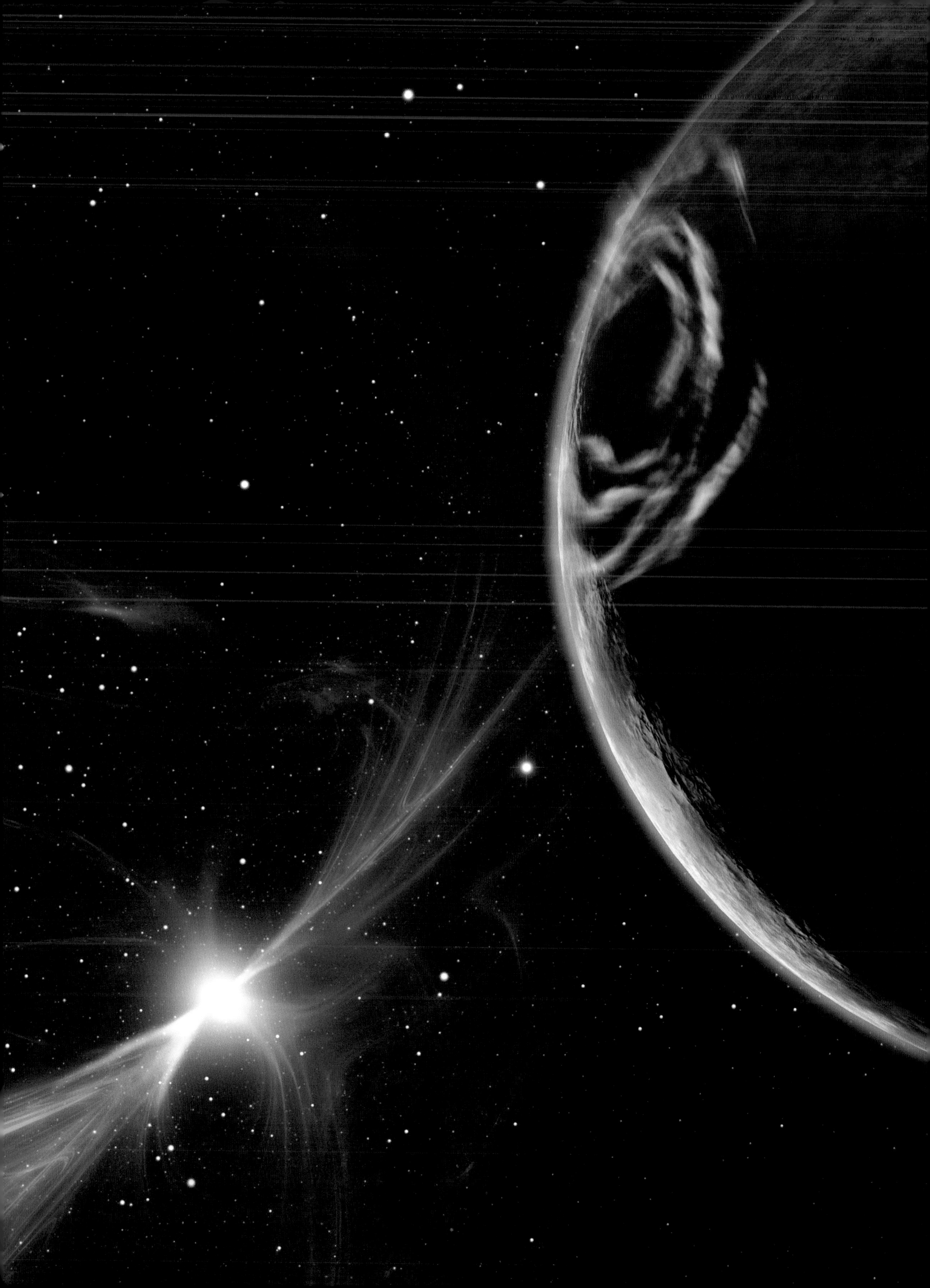

51 PEGASI b

Discovery of a "Hot Jupiter"

The first discovery of a planet orbiting a Sun-like star was reported in 1995. Oscillations in the star's radial velocity over four-day cycles were so large (exceeding 110 meters per second, or 250 miles per hour) that they could easily have been discovered in previous decades. The perturbing planet, 51 Pegasi b, was immediately verified and presaged a flood of similar discoveries in the years to come.

The host star, 51 Pegasi, is a yellow dwarf star only slightly bigger and brighter than the Sun and is located 51 light years away in the constellation Pegasus the Winged Horse. It is slightly more enriched with heavy elements than the Sun and is estimated to be six billion years old—give or take a couple billion. Its present position in the solar neighborhood closely marks the outermost extreme of its Galactic orbit. It comes within 19,000 light years of the Galactic center at the innermost extreme.

The only planet discovered in the system so far, 51 Pegasi b is a gas giant at least half as massive as Jupiter. It is carried around its parent star every four Earth-days in a tight circular orbit barely 0.05 AU away—several times closer than Mercury orbits around our own Sun. Temperatures of this planet's upper atmosphere may hover in the range of 1,800° F. Inflated by the high heat, the atmosphere likely extends to a radius larger than that of Jupiter even though the planet is less massive. The even-hotter layers of gas below may possibly be hot enough to cast a visibly reddish glow.

The discovery of a massive gas giant orbiting a star so closely came as a surprise to planetary scientists. Theories of planet formation had given rise to expectations that the configuration of our own solar system would be typical—rocky terrestrial worlds in the inner planetary region (viz., Mercury, Venus, Earth, and Mars), and gas giants in the outer region (viz., Jupiter, Saturn, Uranus and Neptune). It was not expected that gas giants could form close to their parent stars, where atmospheres of light elements are prone to boil away and get blown off by the stellar wind.

One of the early speculations to account for the existence of 51 Pegasi b was that it perhaps formed much further away from the star and migrated inward only more recently—perhaps due to the combined perturbations of undiscovered planetary bodies elsewhere in the system or planets long-since departed. It is surely possible. The dynamics of planetary systems can indeed be chaotic enough to unravel spontaneously and unpredictably.

On the other hand, maybe the planet did form in the inner region where we now find it while 51 Pegasi was still a *protostar*—or perhaps more rightly: a "double protostar." This secondary, primordial companion may have even become as massive as a brown dwarf (see page 16) before the primary star ignited. But now, after having its atmosphere boiled off for billion of years, there is only a half-Jupiter left over.

No matter its origins, a great many similar examples of exoplanets like 51 Pegasi b have since been discovered. Astronomers have dubbed them "hot Jupiters."

In the scene depicted here, the orb of 51 Pegasi b fills our view from just 150,000 miles away. Its parent star is directly behind us. The faint glow of diffuse gases boiling off its atmosphere trails in the wake of its rapid orbital motion. At the bottom center of the portrait our humble Sun is a sixth-magnitude speck. To the upper right, the stars of Ursa Major shine more conspicuously. The Big Dipper asterism is distorted, but it is not untraceable (as revealed in the chart).

Let us now explore another star system nearby.

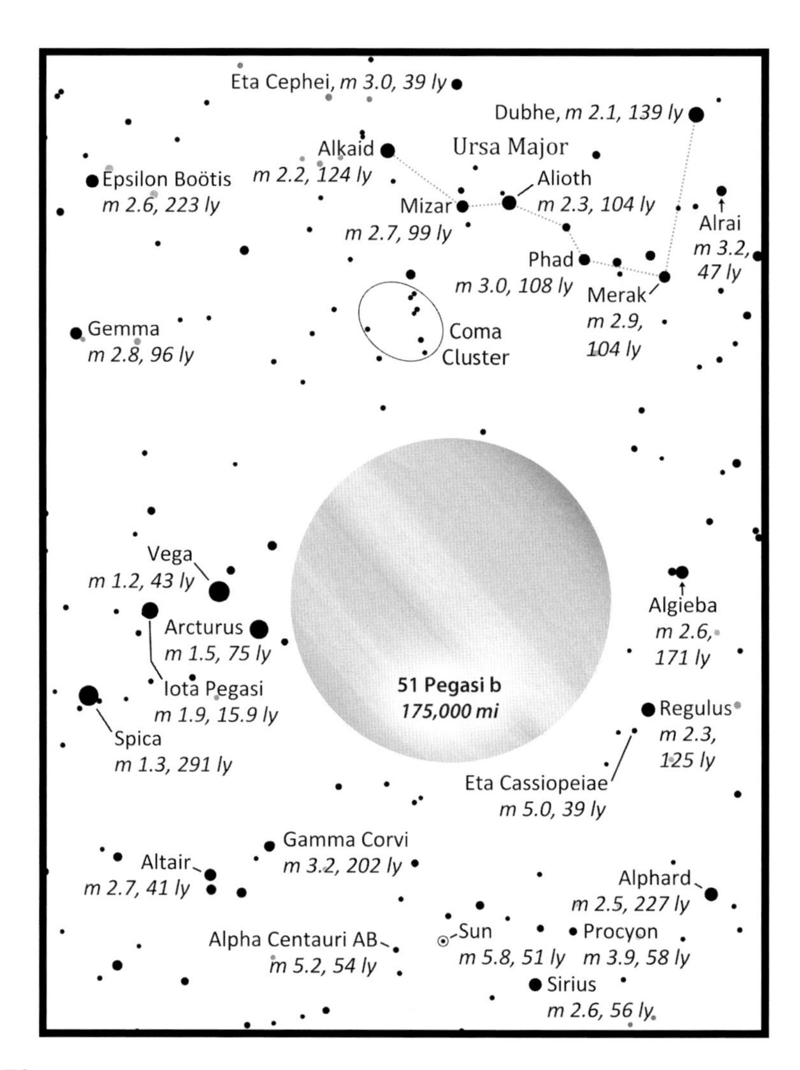

Upsilon Andromedae

A Chaotic Star System

In the panorama of Earthly constellations, Andromeda is Pegasus' next door neighbor. A spaceflight of exactly 10 parsecs (32.6 light years) takes us from 51 Pegasi to Upsilon Andromedae.

Upsilon Andromedae A, the primary star of this system, is slightly more massive, more luminous, and perhaps a billion years younger than the Sun. It hosts four planets that we know of to date. The innermost planet, Upsilon Andromedae b, is another hot Jupiter. Like 51 Pegasi b, its year lasts only four Earth-days. Upsilon Andromedae c and d are two more gas giants located further out; these planets are several times as massive as Jupiter and revolve around their parent star in significantly eccentric orbits. The outermost known planet is Upsilon Andromedae e, which has been called "Jupiter's twin" because it is nearly equivalent to Jupiter in mass and has a similarly sized orbit (5.25 AU). The star also harbors a recently-discovered faint red dwarf—Upsilon Andromedae B. This small stellar companion orbits the primary at a great distance. The size and length of the companion star's orbit remains unknown.

Upsilon Andromedae A is recognized as the first Sun-like star discovered to have multiple planets, when c and d were found in 1999 from radial velocity data.

In the subsequent decade the star was again observed with the Hubble Space Telescope, which supplemented the radial velocity data with astrometric measurements. These latter measurements (which reveal the kind of side-to-side motion that astronomers hoped to detect in nearby low-mass stars in the mid-20th century—see page 64) are two-dimensional. The new Hubble observations allowed astronomers to determine the orbital inclinations of c and d and to refine their masses to approximately 10 times and 14 times that of Jupiter, respectively.

Curiously, the orbits of the eccentric gas giants c and d turned out to be steeply inclined with respect to each other—a difference of 30 degrees. By contrast, the orbital inclinations between the major planets in our solar system differ only modestly—by just a few degrees.

This discovery presented a new puzzle for planetary scientists. Because the formation process of planetary systems is thought to condense out of a disk of spinning material, the resulting configuration of planets is expected to end up more or less *coplanar* (i.e., in a common orbital plane). What, then, could account for the wild condition of Upsilon Andromedae c and d? One explanation is that the planets originally formed in normal coplanar orbits and were subsequently disrupted by other massive bodies—either former members of the planetary system that have since been ejected, or by close approaches of the binary companion star, Upsilon Andromedae B. In any case, c and d may not be in stable orbits even today. So visceral is their gravitational tug-of-war that their orbital parameters are thought to be evolving rapidly. Perhaps one will eventually defeat the other by throwing his opponent out of the arena.

In the scene depicted here we witness the cosmic wrestling match at a great distance—above a polar region of Jupiter's twin, Upsilon Andromedae e. The perennial storm of a polar vortex stirs up the clouds, setting off sporadic flashes of lightning.

Directly above the planet, we can easily locate Upsilon Andromedae's three other known planets. We also happen to see another solar-like star, Gliese 67, less than three light years away (near planet d).

Next we shall explore a star system that is better behaved.

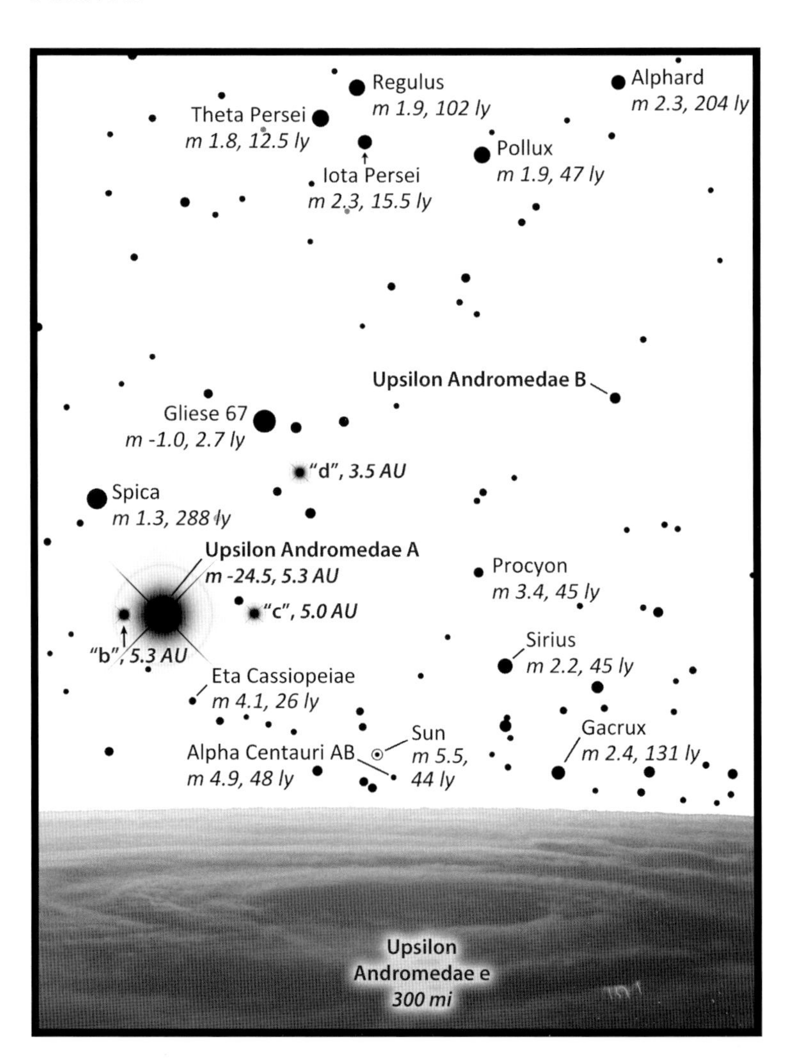

HD 28185 b

A Temperate Gas Giant

As the 20th century drew to a close the rate of exoplanet discovery picked up great speed. Telescopes with larger apertures and high-resolution spectrometers empowered astronomers to monitor fainter stars. HD 28185, for instance, is a Sun-like star in the constellation Eridanus the River that shines at magnitude 7.8 in the skies of Earth from 138 light years away. It is rich in heavy elements and its age is estimated to be ten billion years old—give or take a few billion. Like Delta Pavonis (page 26), it is nearing the end of its life as a main sequence star as it enters into a more luminous subgiant stage. And, like 51 Pegasi, it spends most of its Galactic orbit significantly closer to Galactic center than in the region of the solar neighborhood where we happen to find it today.

The detection of a planet, HD 28185 b, at least 5.7 times as massive as Jupiter, was reported in 2001. It remains the only known planet in the system to date. The orbit of HD 28185 b around its parent star is somewhat Earth-like. It completes its circuit once every 383 days at a distance of 1 AU. The eccentricity of its orbit, however, is more pronounced—comparable to that of Mars.

The orbital distance of HD 28185 b may allow Earth-like temperatures to exist in the planet's upper atmosphere. And if, like Jupiter, the planet itself is orbited by large moons, these worlds too may be temperate.

A moon having the mass of Earth is within the realm of possibility. We may even imagine it to be enveloped with oceans of liquid water. In the scenario presented here, HD 28185 b is portrayed from just such a wet, rocky world a million miles away. The apparitions of three other moons are aligned in a common orbital plane (designated as I, II, and III in the chart). We catch the innermost moon as it transits the dark side of the half-illuminated gas giant.

In the sky of HD 28185 the Sun is too distant to see with the naked eye. Sirius, a few degrees away, is a more conspicuous marker of the Sun's general location. The bright Eridanus star, Sceptrum, looms in the celestial foreground. Closer to the horizon, the constellation Scorpius the Scorpion clambers above the clouds; his claws appear to gape more widely than how we see them from Earth, seeming to convey a more aggressive stance.

Habitable moons are a staple of science fiction. Will actual examples be found someday? A considerable obstacle to their actual existence is the problem of *tidal locking*. The Earth's Moon, for instance, always faces Earth from the same side because its period of rotation (its *spin*) takes the same amount of time as its period of revolution (its monthly orbit around the Earth). This is no mere coincidence, but rather a direct consequence of the Moon's proximity to the Earth's much stronger gravitational field. Hence, the Moon's rotation period is *tidally locked* so that one lunar day lasts an entire month. The temperature difference between midday and midnight on the Moon is therefore quite extreme. If our Moon had an insulating atmosphere similar to Earth's, that would furnish some slight temperature-mitigating effects, but its surface would then regularly endure storms of devastating winds.

The placid scene depicted here in the environs of HD 28185 b suggests special circumstances. It can only be possible if our hypothetical moon resides at an orbital distance far enough away to have skirted the fate of tidal locking. Not many worlds may be so fortunate, as we shall see next.

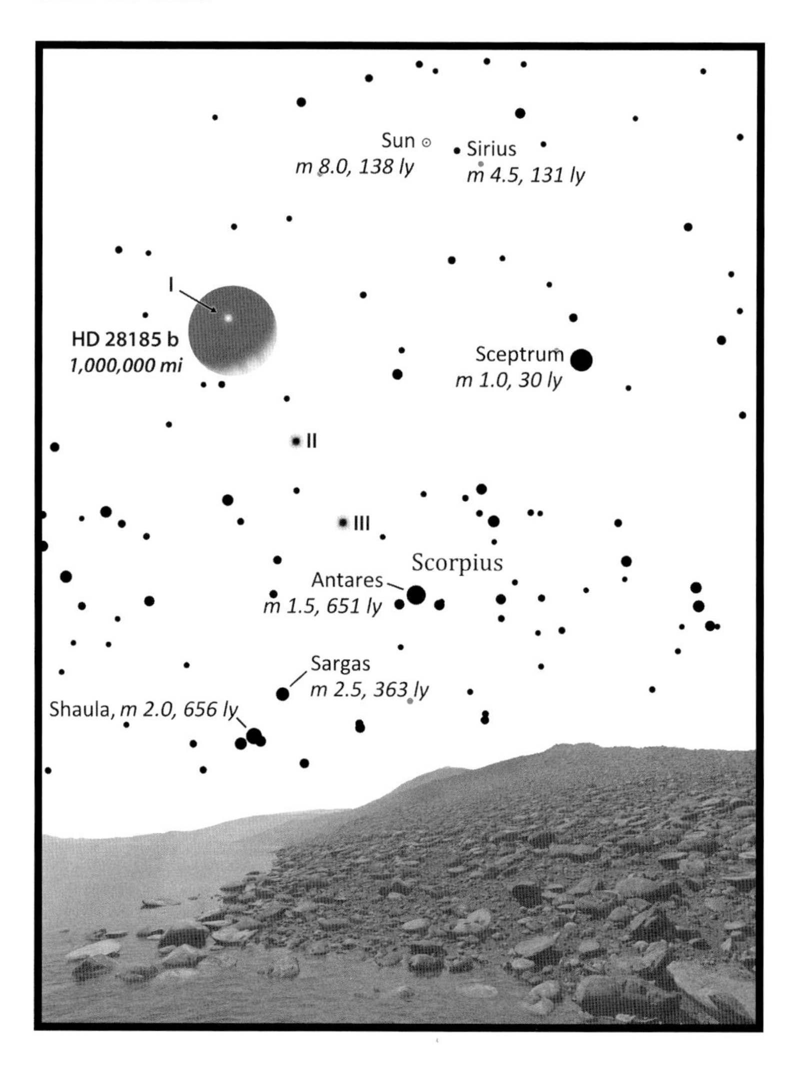

55 CANCRI e

A Transiting "Super-Earth"

Rho[1] Cancri, better known as 55 Cancri, is another Sun-like star. It is located 40 light years away in the zodiacal constellation Cancer the Crab. Like the two previous stars we visited, its chemical composition is rich with heavy elements. It is cooler and older than the Sun and is likely to be in a state transition from a dwarf to a subgiant. The star has a red dwarf companion, 55 Cancri B, separated from the primary star by at least 1,000 AU.

Five planets are known to orbit the star, all discovered through radial velocity oscillations. The earliest one found, in 1996, was 55 Cancri b—another hot Jupiter at an orbital distance of 0.11 AU. More planets were discovered in 2004. Three planets, c, d, and f, each having masses in a range comparable to Saturn, were found orbiting beyond b, within 0.8 AU of the parent star and 55 Cancri e was found to be the closest planet of them all. It orbits the star every 17½ hours at a scant distance of just 1.5 million miles away—a distance barely three times larger than the radius of the star itself. The mass of 55 Cancri e is nearly eight times that of the Earth, hence it is an example of a planetary category which has no example in our own solar system: a "super-Earth."

The close proximity of 55 Cancri e to its parent star, and its rapid orbital period, also make it an opportune target for recording transits—occultations of the star by the planet whenever it crosses in front of the star. These crossings can be detected on Earth as periodic reductions in the star's brightness. If a dip in brightness can be precisely measured, then the size of the planet relative to the size of the star can be determined.

In 2011, two independent teams of astronomers observed transits of 55 Cancri e with space-borne orbiting telescopes—NASA's Spitzer Space Telescope (which operates in infrared light) and the Canadian Space Agency's MOST telescope (operating in visible light). The initial estimates of the planet's diameter yielded discrepant results, but subsequent observations and a reanalysis ultimately converged on a figure near 2.2 times the width of the Earth.

Eight Earth masses packed into a planet in this size-range means that the object cannot be a small gas giant comparable to Uranus or Neptune. The planet must be solid, i.e., a *terrestrial* super-Earth. But the planet is in no way Earth-like. Surely 55 Cancri e is tidally locked, and the temperature of the planet's sunward-facing surface has been estimated to exceed 3,000°F. Gravitational interactions with the nearby star and planet b may also induce the same sort of tidal friction in the planet's interior that is responsible for the global volcanic activity seen on Jupiter's moon Io.

In 2012, new observations with the Spitzer telescope recorded how much infrared radiation emanates from the planet itself as it circles behind the star. This study confirmed the planet's high temperature and provided some new clues about its surface brightness, atmospheric density, and atmospheric composition.

In the scene depicted here we come to within 35,000 miles of 55 Cancri e. The transit in progress is, for us, a brief eclipse. In the background we find the Sun shining at fifth magnitude in the zodiacal constellation directly opposite of Cancer: Capricorn the Goat. The other planets in the system are shining brightly nearby.

But we shan't linger long here, lest we melt. Let us move on to a cooler location.

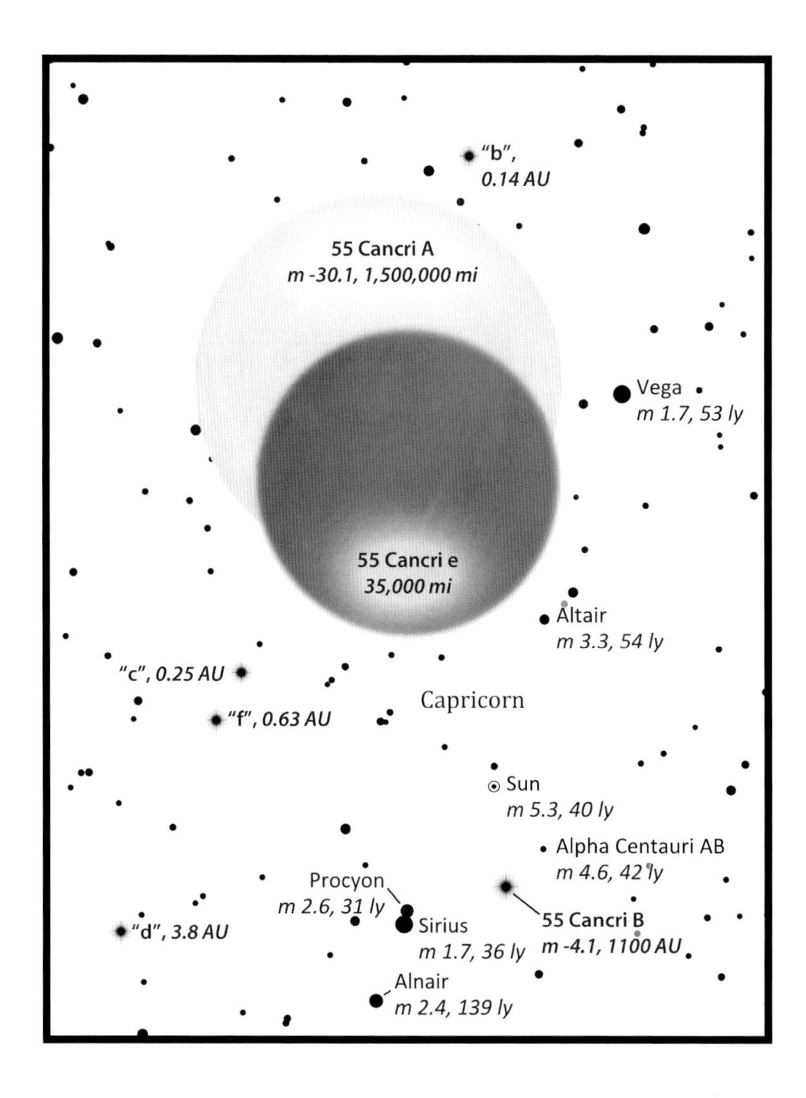

GLIESE 581

Are Planets "Habitable" in Red Dwarf Systems?

In the skies of Earth, the red dwarf star Gliese 581 shines at magnitude 10.6 from 20 light years away in the zodiacal constellation Libra the Scales. Its present Galactic location is near the outermost extreme of its moderately eccentric Galactic orbit, yet it is somewhat deficient in heavy elements compared to the Sun. Being a red dwarf, the star's age is difficult to estimate.

To date, four confirmed planets are known to orbit the star—all of them super-Earths within 0.22 AU of the parent star. Two more super-Earths have been proposed to exist at 0.14 and 0.76 AU, but this interpretation was immediately challenged and subsequent observations do not support their existence.

Nevertheless, the proposed planet at 0.14 AU, Gliese 581 g, garnered a great deal of public attention when it was announced in 2010. At this distance from the cool red dwarf, a solid planet could be temperate enough to allow for the existence of liquid water. It is a possibility that unfailingly arouses a recurring theme of speculation: could a planet so situated be suitable for life?

The question demands that we revisit the problem of tidal locking that we previously considered on our sojourn to HD 28185 b. A planet that has formed very near its host star would surely become tidally locked. Yet this is where a planet *must be* to sustain temperatures above freezing when the star is a low-energy red dwarf.

If a planet's rotation matches its orbital period, it is deemed to be in a 1:1 *spin-orbit resonance.* The same side of the planet would face the star unflinchingly. Other spin-orbit resonances are possible. Mercury, for example, rotates once every 59 days as it revolves around our Sun every 88 days. This outcome is deemed a 3:2 resonance because the planet rotates three times in the same amount of time that it completes two orbits (this curious relationship may be due to the eccentricity of Mercury's orbit, which results in non-uniform orbital speeds).

If Gliese 581 g actually exists, its orbital eccentricity is negligible and a spin-orbit resonance of 1:1 would therefore be expected. If, like the Moon, it lacks an atmosphere, its unlit hemisphere would be as cold as deep space. If, on the other hand, it possesses a thick atmosphere like Venus, worldwide temperatures would be uniform, but at unbearable levels of heat and atmospheric pressure. One might hope for the planet to have an atmospheric density in between the two extremes, comparable to the Earth, but that would be a precarious ideal for a planet locked 1:1. One side of the planet, cloaked in perpetual night, would be cold enough to freeze the air into solid particles. An entire hemisphere would be coated with an irrecoverable pile of snow. Meanwhile, on the daylight side, relentless heat would stimulate ocean evaporation and outgassing of carbon dioxide from uncovered rocks. One process tends toward a lunar destiny, the other process tends toward a Venusian one.

In the scene portrayed here the lunar destiny has ultimately prevailed on Gliese 581 g. We take in an aerial view of the terrain from a location not far from the planet's terminator—the dividing line between night and day. In the foreground, stratified outcrops demark ancient courses of torrents that had once fed an insatiable sea. An endless desert lies beyond where there once had lain an ocean floor. Gliese 581 glares mercilessly near the horizon. It just hovers there; it never sets. The wispy remnants of an atmosphere glimmer in the perpetual twilight, yielding readily to the penetrating starlight.

Let us now return to the stars.

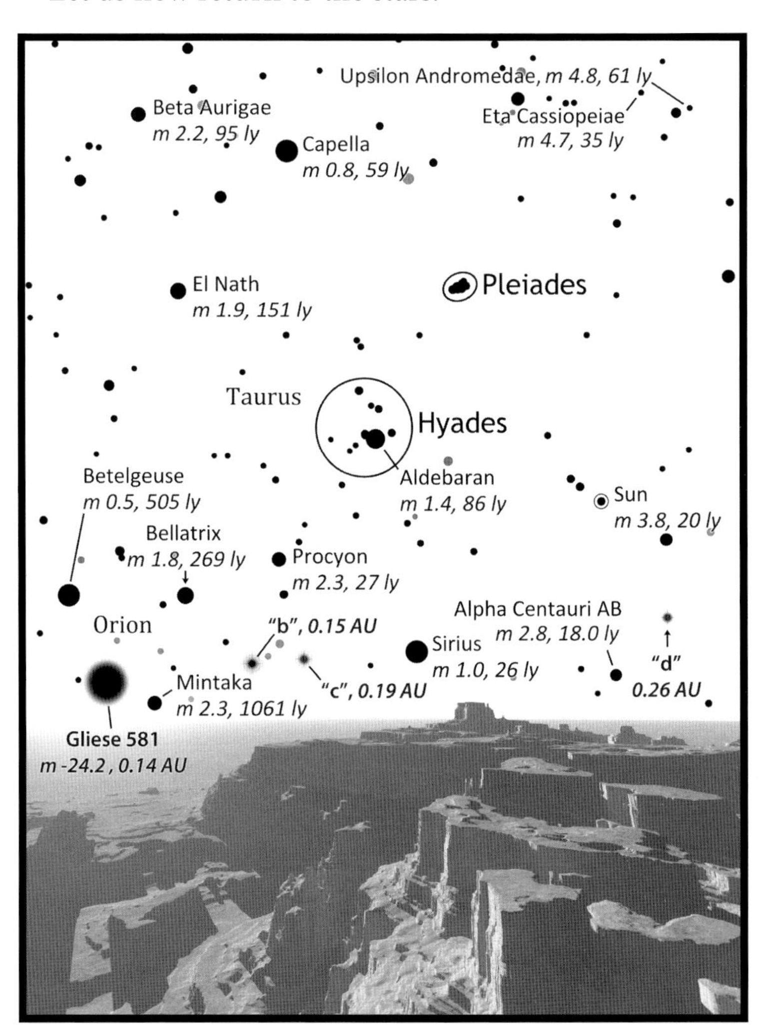

HD 73526 b AND c

Counter-Revolving Planets

HD 73526 is a Sun-like star residing in the southern constellation Vela. It shines at magnitude 9 from its location 324 light years away. The star is estimated to be twice as old as our Sun (at 9.2 billion years, give or take 1.5 billion years), it is richer in heavy elements, and its Galactic orbit is significantly eccentric (30%). Its distance from the Galactic center, which varies between 19,000 and 37,000 light years, is presently increasing.

The star is known to possess two planets, b and c, each having at least twice the mass of Jupiter. They were discovered in 2002 and 2006 respectively, at orbital distances corresponding to the Venus-Sun distance and the Earth-Sun distance. Their orbital periods (approximately 188 and 377 days) exhibit a 2:1 ratio. Every time b goes around twice, c goes around once. Hence, like Gliese 876 (page 20), HD 73526 is another example of a system that manifests an orbital resonance.

Curiously, though, astronomers later calculated that the planet-to-planet gravitational interactions, which must occur between these closely spaced titans, do not lend themselves to long-term stability. Their orbital configuration becomes rather improbable under normal assumptions. But orbital simulations also show that the system *could* be stable with a daring presumption: that the planets orbit the star in opposite directions! This novel solution not only satisfies theoretical orbital simulations, but it also gives the best match to observed nuances in the timing cycles of the radial velocity data.

However, one theoretical problem still remains: how do we reconcile counter-revolving planets with natural circumstances? Planets, we assume, are formed out of rotating disks of material (see page 14). It is implausible that a protoplanetary disk could rotate in two directions at once! All planets generated from a common platter of material ought to be revolving in a common direction, which should also match their host star's direction of rotation.

By 2010, though, it was becoming clear that exceptions abound. For planets in orbits that happen to exhibit transits (e.g., 55 Cancri e), it is possible to derive their orbital inclinations relative to the spin axes of their host stars. This is made possible through spectroscopy by measuring changes in the breadths of spectral lines as the planet transits across the width of the rotating star (the *Rossiter–McLaughlin effect*). It turns out that, among all the transiting hot Jupiters, a high proportion of them are found in orbits steeply inclined to their host star's equator (which harkens back to the case of Upsilon Andromedae; page 72). Even more surprising is that a significant minority of the sample systems were found to have planets in retrograde orbits.

What could possibly explain their backward ways? Hypotheses abound. They might have been free-floating planets captured by stars from interstellar space. Or they may be the outcome of chaotic interactions with undetected planets (some of which may have since been ejected into interstellar space).

In any case, we are entitled to marvel at HD 73526. In the portrait presented here, our imaginary photographer has opted for a multiple exposure image to capture a sense of the contrary motions. HD 73526 c, nearby, drifts by us from right to left, while b, in the distance, saunters in the opposite direction.

Next, we'll visit a six-pack of transiting exoplanets.

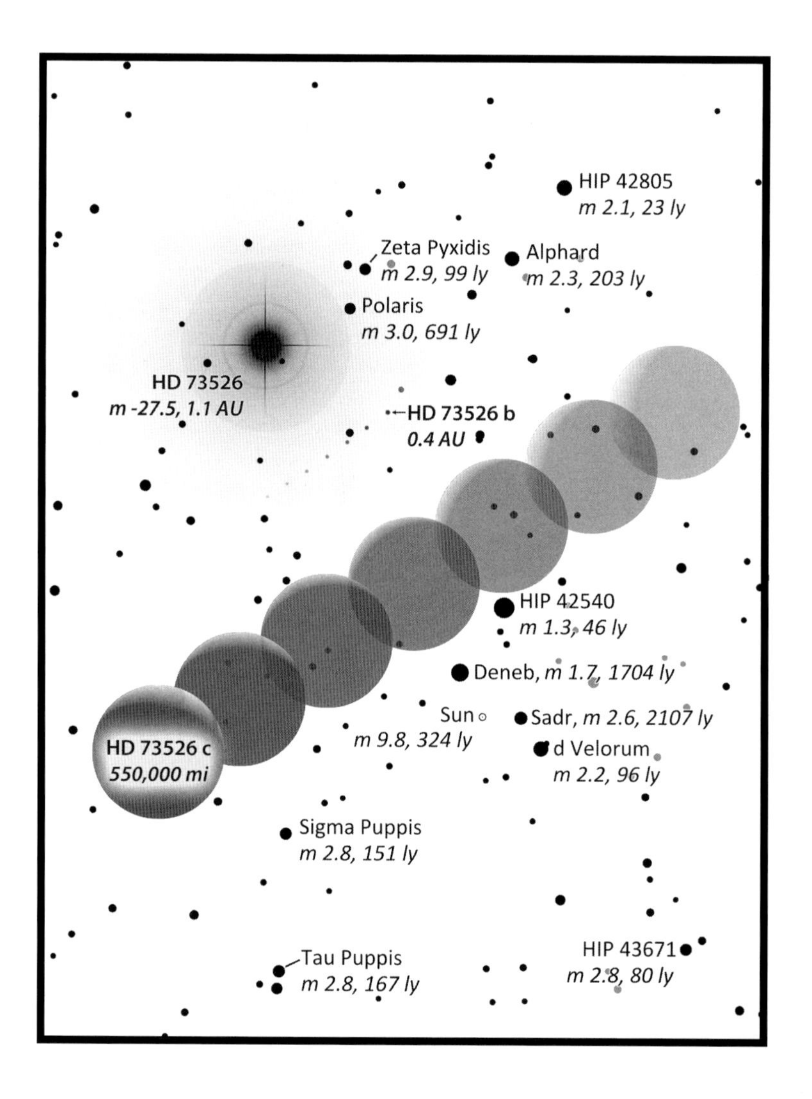

Kepler-11

A Compact Star System

Early in 2009 a space observatory built by NASA was launched into Earth orbit. Named "Kepler," in honor of the renowned German astronomer (see page 50), its mission would be to keep nearly 150,000 stars in and around the constellation Cygnus under constant surveillance for at least 3½ years. If, at any time and for any reason, any one of those stars changed in brightness even slightly, the occasion would not escape Kepler's vigilance. The intention of Kepler's designers was to record the transits of exoplanets across the faces of their parent stars. Analysis of the data recorded so far have already revealed thousands of signs of potential planets.

Catching a transiting exoplanet requires a great deal of luck. The planet's orbital plane must be in near-perfect alignment with the Earth—otherwise, no transit will ever be observed. And when transits do occur, they don't last long—just minutes or hours, depending on a planet's orbital speed. Kepler overcomes these daunting factors of chance by brute force: i.e., the sheer multitude of stars that it observes simultaneously and continuously. This ensures that at least *some* proportion of them will surrender their secrets.

One of the most curious star systems that was revealed among the mission's initial batch of planetary candidates is Kepler-11. The 14th magnitude star exhibits a spectral type similar to the Sun. No parallax data exists, so its distance and age is not yet well-determined, but it is estimated to be older than the Sun and roughly 2,000 light years away.

Kepler-11 is known to have six transiting planets ranging in size between two and four-and-a-half times that of the Earth. The five innermost planets, b through f, all orbit the star closer than Mercury orbits our own Sun. The sixth planet, g, revolves around the star at an orbital distance comparable to Venus. This tight configuration makes the Kepler-11 system the most compact star system yet discovered.

The estimated masses of the five inner planets (determined by subtle changes in the timings of the orbital periods, caused by planet-to-planet gravitational perturbations) make them each less dense than the Earth. Hence, we may suppose that they are composed of light, gaseous elements—not unlike the outer planets of our own solar system. But, as with the examples of hot Jupiters (e.g., page 70), the presence of low-density gas-planets in the inner planetary region provokes a mystery. It remains unknown whether their formation occurred farther away from the star and they have since spiraled inward from their original location, or if they formed where we find them now and much of their original mass has since evaporated into space.

In the scene depicted here we opt for a telephoto view to capture the Kepler-11 system. We catch the innermost planets, b, c, and d, in a simultaneous triple-transit. Thin crescents of the illuminated sides of planets e and f can be spotted to the upper left and of planet g nearby to the lower right.

In line with the plane of planets resides the Sun. Though unseen in the portrait, its location marked on the chart where it resides as a 14th magnitude object.

Let us next examine another Kepler discovery.

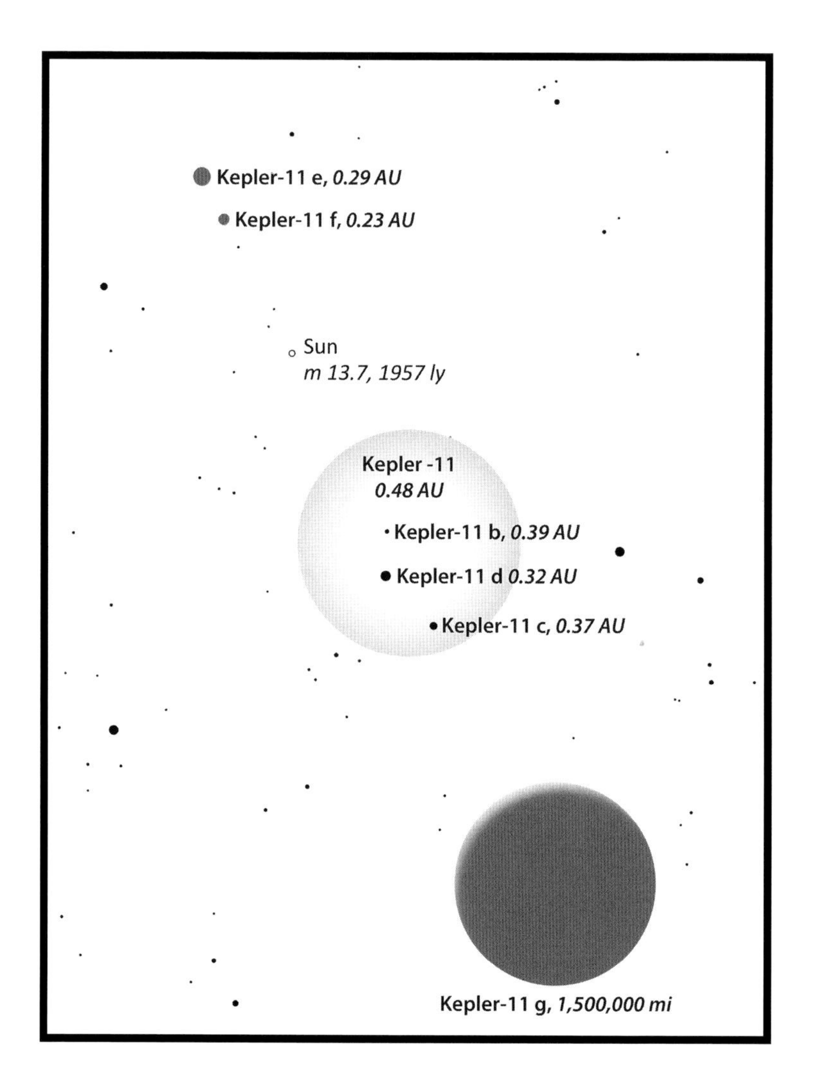

KEPLER-16 b

A Transiting Circumbinary Planet

So far in our journeys in space and time we have encountered a few examples of multiple-star systems, such as Alpha Centauri, Sirius, Eta Cassiopeia, and Albireo. Multiple-star systems are in fact common throughout the Galaxy. Half of all existing star systems may be binary stars or even contain three or more stars. Some binary systems are now known to have planets, like two examples we've recently visited: Alrai and Upsilon Andromedae. Now we shall visit another planet detected early on by the Kepler mission, which orbits not just merely one component of a binary system, but both stars at once.

Kepler-16 A is an orange dwarf star slightly less massive and less luminous than Epsilon Indi. Its red dwarf companion, Kepler-16 B, is one of the least massive stars ever discovered. It orbits the primary component every 41 days in a moderately eccentric orbit 0.2 AU away.

The system, which lies approximately 200 light years away from the Sun in the constellation Cygnus the Swan, is a perfect example of an *eclipsing binary.* Because the system's orbital plane is coincidentally aligned with the location of the solar system, the two component stars undergo mutual eclipses when seen from Earth. At regular intervals the A component will move directly in front of B and vice versa. Each eclipse dims the amount of starlight reaching Earth—hence, an eclipsing binary is also an example of a *variable star.* The most famous eclipsing binary in the heavens is the bright star Algol—meaning "the ghoul" or "demon star"—in the constellation Perseus. Algol's variability has been well known since ancient times.

In the case of Kepler-16, its binary nature was made obvious by Kepler spacecraft. The system's regular eclipses were recorded like heartbeats. Curiously though, a dip in brightness was also recorded between the times that the two stars eclipsed each other. This betrayed the existence of a third body—Kepler-16 b.

Kepler-16 b turns out to be a planet slightly smaller and denser than Saturn. It orbits the AB pair at a distance of 0.7 AU, making it the first known *circumbinary* planet ever discovered—i.e., a planet having an orbit that encircles a binary pair.

In this scene we have a look at the Kepler-16 system from within. We take our position on a hypothetical moon orbiting the gas giant. The binary host stars, Kepler-16 A and B, shine through the gaps of a hypothetical ring-system encircling the planet. The orange primary star, Kepler-16 A, appears thirteen times fainter than the Sun does from Earth. The red dwarf secondary is 1,000 times fainter than A. With so little sunlight trickling in, Kepler-16 b must be an icy world.

Since this is one of the closest target stars in the Kepler program, the view from Kepler-16 brings us back within the range of many stars familiar to stargazers on Earth. The seven brightest stars of Orion are all visible in the upper right corner, though somewhat scrambled. Saiph and Bellatrix lie in a nearly straight line with Kepler-16, as do the "belt" stars Alnilam and Mintaka, so that they form up bright and tight *visual doubles.* To the upper left, Kochab and Dubhe (both of orange hue and both corner stars of the bowls of our Little and Big Dippers respectively) are also seen paired together.

Next we shall set out to reach our final destination for this itinerary. In order to get there, we must venture the furthest distance yet from home.

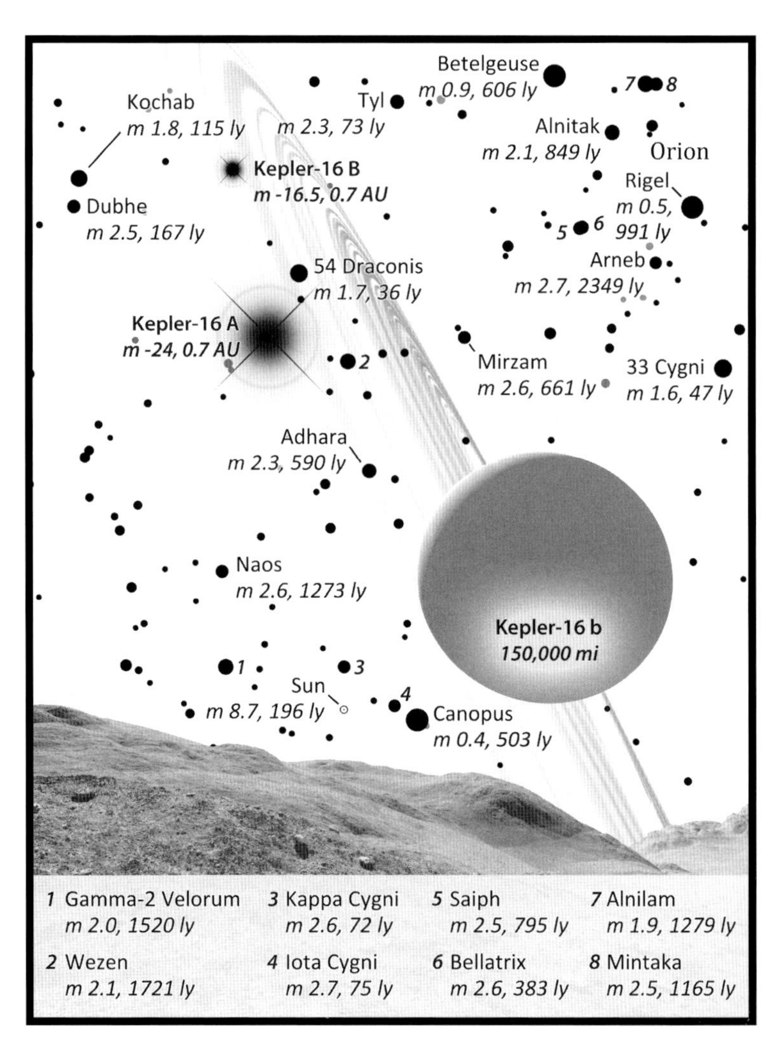

HIP 13044 b

Immigrant Planet from Another Galaxy

Our travels thus far have been mostly confined to the plane of the Galaxy. We did step out quite far to visit PSR B1257+12 (page 68). We now must venture out even further to reach our next destination, a 10th magnitude star in the southern constellation of Fornax the Oven, cataloged by Hipparcos (page 8) as HIP 13044. The star is too far away and too faint for Hipparcos to derive a reliable parallax measurement, but its distance is estimated to be in the range of 2,300 light years.

We find the star at this odd location because its Galactic orbit is steeply inclined to the Galactic plane. Intriguingly, it is just one member of a group of stars in similarly inclined orbits. Astronomers reckon that these stars are the remnants of a dwarf galaxy that merged with the Milky Way Galaxy several billion years ago. Because the mass of the Milky Way was superior, the smaller galaxy was ripped to shreds. The stars that formerly comprised the dwarf galaxy now encircle our Galaxy in a *tidal stream.* Hence, HIP 13044 is not a native citizen of the Milky Way—it is an "immigrant star."

The physical characteristics of HIP 13044 place it at a late stage of stellar evolution that astronomers term *the horizontal branch.* Normal stars shine by fusing hydrogen into helium. When a star with a mass similar to the Sun reaches about ten billion years in age, it begins to run out of hydrogen fuel. The accumulated helium sinks to its core and fusion in the outer shell accelerates. The star then balloons into a red giant. Eventually the inner reservoir of inactive helium accumulates so much mass that the core collapses under the weight of its own gravity. The core temperature rapidly escalates, setting off an internal explosion called a *helium flash.* This event radically reorganizes the star's structure. The core stabilizes itself by fusing its helium into heavier elements. The surrounding shell also contracts and continues fusing its hydrogen at a higher temperature. The outward appearance of the former red giant is thereby transformed into a smaller, whiter star—a *horizontal branch* star. It will remain as such for another billion years or two until all of its atomic fuel is exhausted, heralding the conclusion of its life.

In 2010 a series of radial velocity measurements of HIP 13044 revealed that it is closely orbited (at 0.12 AU) by a hot Jupiter—HIP 13044 b. We cannot suppose that the planet has always been this close to the star because it would not have survived in its present location during its host star's preceding red giant phase. It must have spiraled inward—either due to friction induced by the former red giant's extended stellar atmosphere or due to gravitational interactions with other planets that may also exist in the system. In any case, HIP 13044 b is thought to be coeval with its parent star and hence resides in the Milky Way as an "immigrant planet." The system is extremely poor in heavy elements, and though the size of HIP 13044 b remains unknown, its composition of super-heated light elements should make it enormous.

We now bring our tour of alien worlds to a close. Though our sampling represents only a small portion of the several hundred confirmed exoplanets discovered to date, we should not fail to be impressed with how radically incomparable other star systems are to our own. No analog to the solar system, once assumed to be prototypical, has yet to be discovered. Nor have we found anywhere a planet resembling Earth. But that does not necessarily mean that our homeworld must be utterly unique.

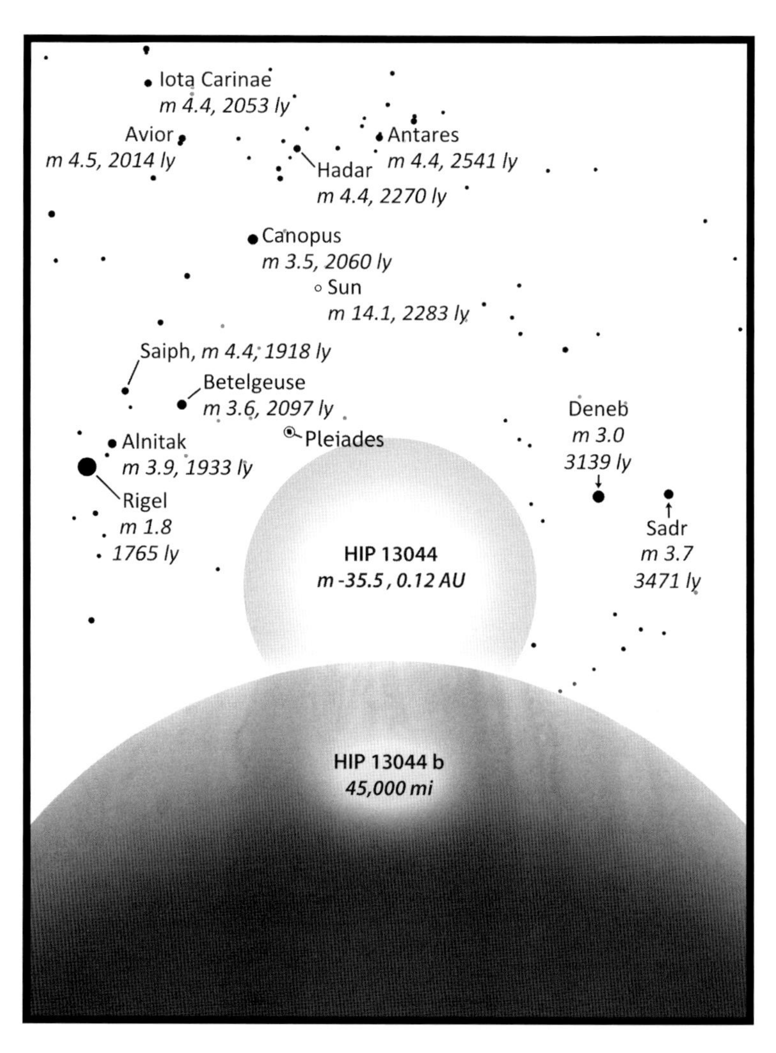

Fourth Itinerary

Hints of Parallel Earths

Rare Earth

What Makes Worlds Like Ours Special?

"Perhaps in 25 years [we could] image a planet around a star within 13 parsecs of Earth, if it exists. A terrestrial-size, Earth-size planet, with a resolution to see oceans, clouds, continents, and mountain ranges. And if that doesn't get your heart pumping, I don't know what will." - *Daniel Goldin, NASA administrator, 1996*

Where in the Galaxy might we find twins of Earth? So far, the main method utilized by astronomers in the quest for planets beyond the solar system—Doppler spectroscopy—has been ill-suited to find planets as small as our tiny planet Earth. An Earth-mass planet simply cannot induce large enough radial velocity oscillations on its host star to be detected in spectrographs.

The *transit* method of planet detection seems to be more promising. An Earth-sized planet observed to cross in front of a Sun-sized star only dims the star's light by a scant ratio of 1:12,000—but the Kepler Mission (page 82) is designed to notice such subtle events. The first Earth-sized exoplanets ever discovered, two of five planets closely orbiting Kepler-20, were announced in 2011. But they cannot be twins of Earth. They have orbits smaller than Mercury's, with periods of 6 and 19½ days respectively. Their tight proximities to their host star make them inhospitable. Will Kepler find an Earth-sized planet in an Earth-like orbit? Kepler's mission may have to be extended several years longer than planned in order to do so. This is because most of its target stars flicker in brightness more than Kepler's engineers anticipated, creating "background noise" in the data. This contingency requires increasing the minimum number of transits it takes to make a detection of a small planet certain. The smallest planet detected in a long-period orbit so far, Kepler-22 b, is also not "Earth's twin"—it is more than double the size of Earth.

Future technologies may even make it possible to image an Earth-sized planet directly. A large enough space-borne telescope could achieve this. The most ambitious plans on today's drawing boards envision arrays of multiple space telescopes that can combine their light while flying in formation so that they effectively function as one colossal instrument. The *Terrestrial Planet Finder* (NASA) and *Darwin* (ESA) are two examples. Both were canceled in 2007 due to lack of budgetary support but eventually the day may come when the means to develop such grand designs are within reach. Future space telescope arrays may even surpass the capabilities of the aforementioned proposals. Indeed, there is no theoretical limit on how big a space-borne telescope array can be. An array spanning several miles could potentially resolve fine features of Earth-like planets in nearby star systems—just as Dan Goldin once envisioned, though not likely before the date (ca. 2021) he predicted.

In the meantime, we can only speculate on where Earth-like planets might be. What clues should we follow? A good starting point would be to recognize, as the Italians say, that "Rome wasn't built in a day." It took billions of years for the Earth to develop into the oasis in space that bore our existence. If "other Earths" do exist elsewhere in the Galaxy, we may likewise expect them to be the outcomes of natural processes operating on similar timescales. Surely not every star system in the Galaxy is suitably configured to allow these processes—many may even prohibit them.

The most essential processes are not difficult to surmise. The Earth formed within the environs of a stable, long-lived star. It condensed into a solid planet with a particular mass, rate of spin, and orbital distance that permit livable surface temperatures. It is gravitationally joined with a large orbiting moon which, through tidal interaction, helps to keep the Earth's spin axis generally perpendicular to the direction of the Sun. This axial orientation keeps seasonal climate changes modest. It is enveloped by oceans of liquid water, originating from volcanic steam and colliding comets. Vast continents interrupt the sea, made possible by an enduring geological activity—*plate tectonics*. The Earth also has the good fortune of being "protected" by a planetary sibling: the gas giant Jupiter, which (some astronomers believe) sweeps the solar system clean of long-period comets. Photosynthesizing organisms could then emerge, producing oxygen that, in time, would saturate the atmosphere. Oxygen, in turn, facilitated the evolution of animal life and biological diversity.

These major ingredients of an Earth-like world are what we shall explore in this last itinerary. We will travel to specific Galactic locations where Earth-like planets could possibly be developing in ways that parallel our own planet's long natural history. Our tour of these "imaginary places in likely spaces" will give us an appreciation of what makes worlds like ours special.

Let us begin at the beginning.

A Star is Born

Rho Ophiuchi Cloud Complex

Deep within a nebula of gas and dust, four hundred light years away from the Sun, a condensed sphere of molecular clouds, surrounded by a flat disk of accreted material, ignites into a blazing ball of nuclear energy. Such is how stars are born. It is an everyday occurrence in our tremendous universe—no different today than it was the day our Sun was born 4.6 billion years ago.

The surrounding disk will likely consolidate into many planets (see page 14). Could any such planet eventually develop into an Earth-like world? Its fate will be determined largely on the characteristics of the star itself, and crucially upon its most essential attribute: the star's *mass.* The more massive a star, the hotter and brighter it burns—and the shorter its life.

The most familiar stars seen by Earthbound stargazers as outlining major constellations tend to be very luminous. Of the 500 brightest stars in the Earth's sky (stars that are brighter than fourth magnitude), only a few are less luminous than the Sun: Epsilon Eridani, Tau Ceti, and Alpha Centauri B—which appear bright to us only because they happen to be nearby. The rest of these are all more luminous than the Sun—mostly because they are more massive, sometimes because they are nearing the end of their stellar lifetimes, and often both. Hence the most conspicuous stars that greet us nightly are not likely to be other suns that bring daylight to other Earths.

The problem with massive stars is that they simply do not live long enough for Earth-like planets to develop. The most massive stars in the Galaxy cannot last much longer than a few million years—thousands of times shorter than the life expectancy of the Sun! Even a star with just twice the mass of the Sun cannot last much longer than one billion years. Only a star packing less than 1.25 times the solar mass can be a normal star for as long as the current 4.6-billion-year age of our Sun.

Conversely, the less massive a star, the cooler and fainter it burns—and the longer its life. One would never suspect from naked-eye stargazing that most stars that surround us in the Galaxy are red dwarfs of spectral type *M.* The brightest example of an M-type dwarf is Gliese 825, which shines at magnitude 6.7 in the skies of Earth from 12.9 light years away—just below the threshold of naked-eye visibility for the typical human observer. Subject the skies to the scrutiny of telescopes, however, and the heavenly complexion changes. Within 50 light years, we find that M-type dwarfs outnumber all other stellar types combined, and we may expect that this preponderance holds true in every region of the Galaxy.

Red dwarfs have fantastic life-expectancies—ranging in the hundreds of billions of years, which is a span of time that far exceeds the present age of the universe. Hence, no crucial time-limit is imposed by red dwarfs upon developing worlds. But red dwarfs do impose other types of difficulties to potentially Earth-like worlds. We have already discussed the problem of tidal locking (pages 74 and 78). Another hazard imposed by red dwarfs are stellar flares. Many red dwarfs are flare stars which can dramatically increase in brightness for short periods of time. Gliese 825 and Proxima Centauri are two such examples. Such sudden and prodigious outbursts of energy do not bode well for planetary habitats nearby.

Depicted here is a newborn star in the Rho Ophiuchi Cloud Complex. If it is to someday harbor Earth-like worlds, a favorable mass is required. A mass closely comparable to that of our Sun is most favorable indeed.

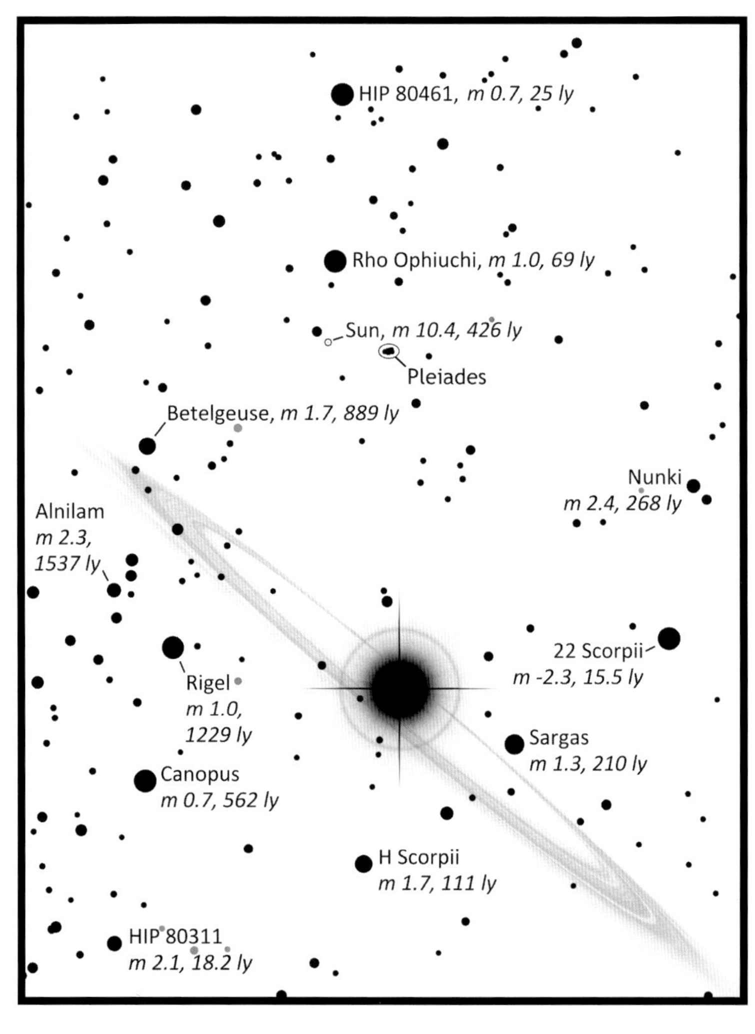

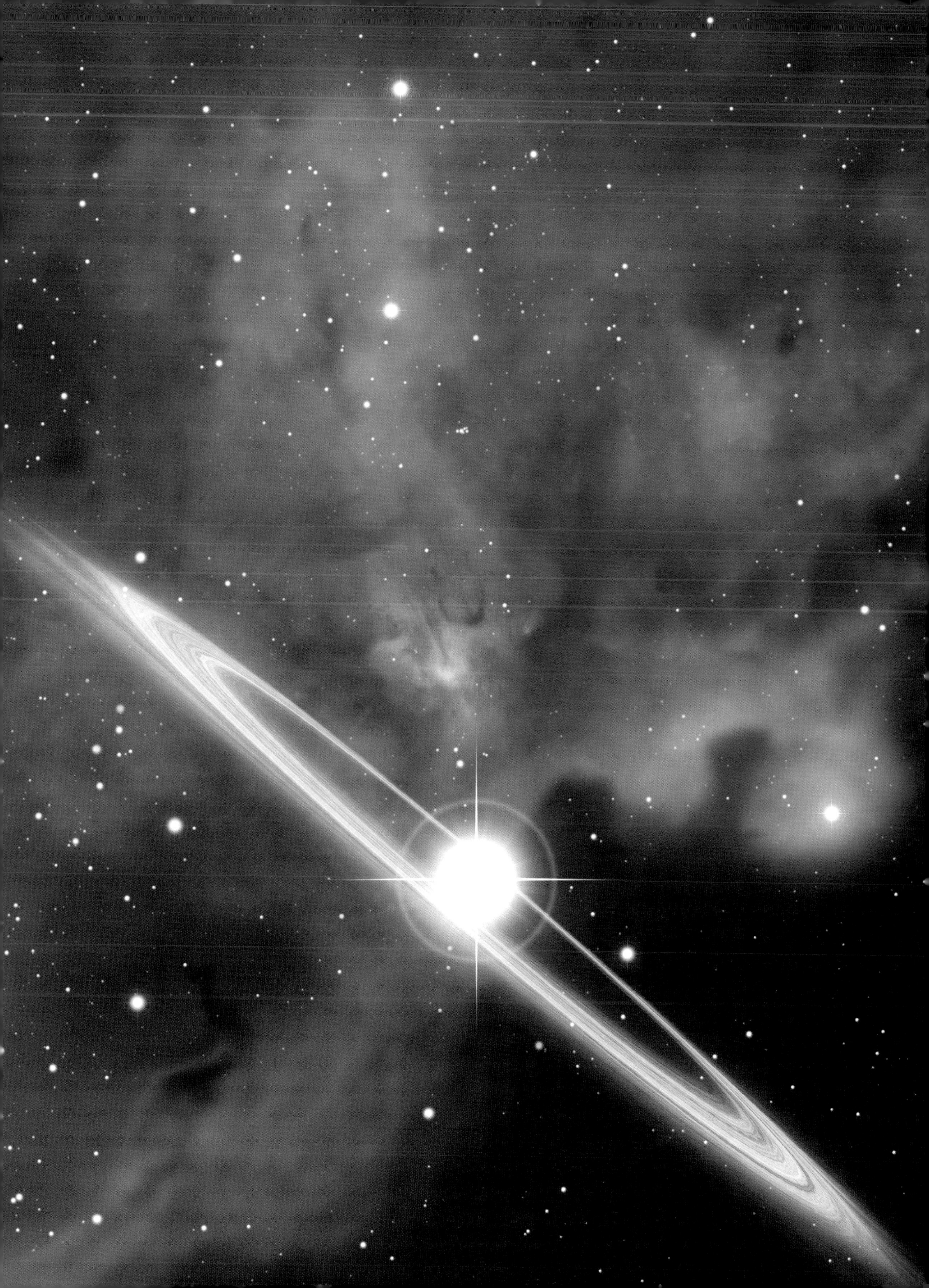

When Worlds Collide

Theta Carinae Cluster

As time goes by, a newborn star's accretion disk tends to consolidate. Originally composed of dust grains, this material may fuse into *planetesimals* (bodies as large as asteroids) at the time of stellar ignition, and these may meld into *protoplanets* (bodies as large as Mars) after a few million years. Protoplanets may continue to grow in size for tens of millions of years, until all the loose material is assimilated.

A star system's inner planetary region may become so crowded with protoplanets that some of them may share common orbital tracks. Planetary scientists theorize that the primordial Earth was once stalked by a Mars-sized object that formed in the same orbit as the Earth's, either 60 degrees ahead of or behind it. Eventually, the co-moving body fell out of place, the orbital gap was closed, and the worlds collided. The resulting impact ejected a swath of debris that condensed into at least two new large bodies, which encircled the Earth. These latter objects reenacted the previous drama, giving chase until they too collided—this time merging into a single satellite that became the Earth's Moon. The radical difference in appearance between the near- and far-side surfaces of the Moon is taken as one line of evidence for this latter collision.

A similar sequence of events may be taking place in primordial star systems throughout the Galaxy. The scene depicted here presents a snapshot of worlds colliding in a Galactic location where it is very likely to be happening right now. We have traveled to the Theta Carinae star cluster—a congregation of young stars (about 50 million years old) easily seen from the southern hemisphere of Earth, 500 light years away. The cluster is also known as the "Southern Pleiades" owing to its resemblance to its more famous northern namesake. Here, we view the cluster from within—about 20 light years away from its center. Theta Carinae appears as bright as Venus and a plethora of other bright cluster members also predominate. Individual designations are mostly omitted in this finder chart for the sake of readability.

Arrayed below us is an aerial view of a hypothetical Earth-sized planet. It is taking shape in a star system near the edge of the cluster. At this early stage of development, we capture the exact moment when two orbiting satellites collide. As these lunar objects meld into a single satellite, they parallel the creation of the Earth's Moon.

The recent impact of the lunar objects' protoplanet progenitor had wiped out the planet's original atmosphere. But a new atmosphere has been gradually building up ever since, evidenced by the luminous streaks of incoming meteors. The planet's surface presents a grim scene of large impact craters protruding above an ocean of magma. The entire landscape is repeatedly heaved up and down by the tremendous tidal forces exerted by the closely orbiting satellite. Over time, these tidal forces will conspire to lift the satellite into a higher orbit while the planet's crust relaxes, cools, and solidifies.

A large moon is essential for stabilizing the axial tilt of Earth-like planets. Without it, planets may be vulnerable to toppling over. This may have actually happened to Mars in its past. The Earth's modest 23½ degree axial tilt gives rise to the seasons: spring, summer, autumn and winter. But if the axis was allowed to get more steeply inclined, the seasons would be more extreme and heralded by profusions of hurricanes. Such conditions would not bode well for the advancement of life!

Next let us reconnoiter a world at a more advanced stage of development.

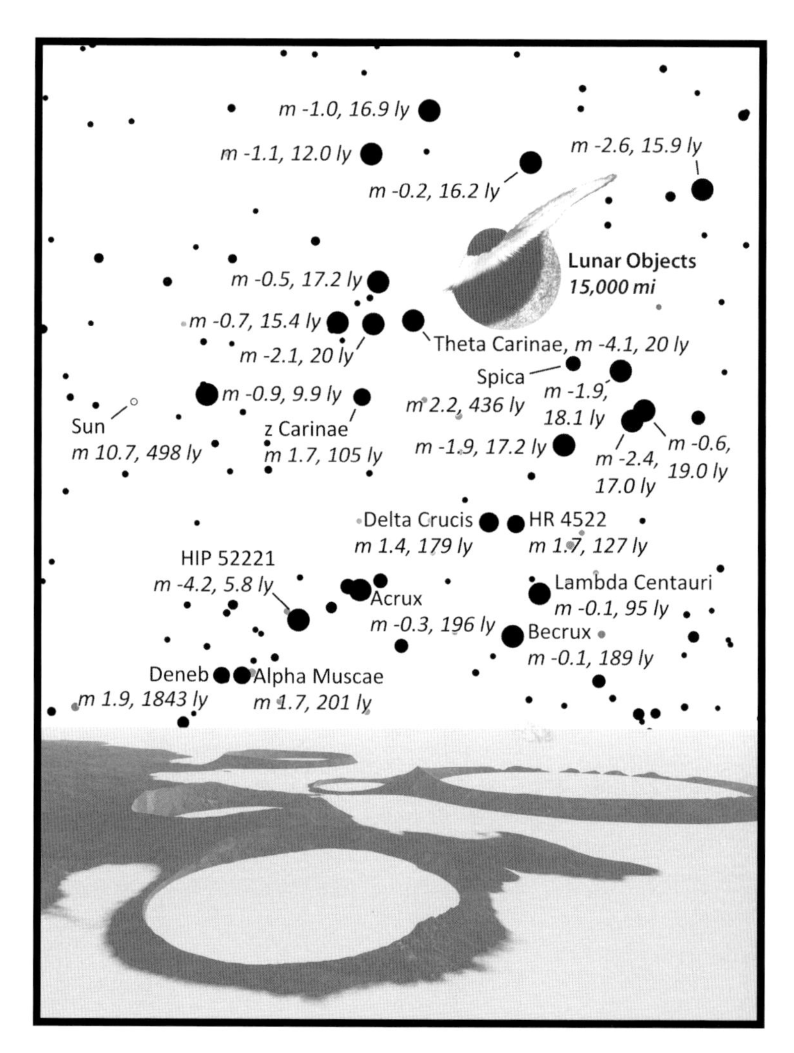

WATER WORLD

Ptolemy's Cluster (M7)

The most magnificent feature of Planet Earth is its abundance of liquid water. Our oceans cover 71 percent of its surface and are two miles deep on average. Where did all this water come from? And why are the other large terrestrial planets of our solar system utter deserts?

Most of the water now in our seas was likely delivered up from the bowels of primordial Earth in the form of steam—seeping up through the rocky crust and billowing forth in volcanic eruptions. These outgassing processes humidified the early atmosphere, creating clouds that surrendered their moisture back down to the Earth's surface in the form of rain.

Earth was also supplemented with water from external sources. Comets, which contain mostly water ice, must have frequently crashed into the primordial Earth. Meteors with more modest water content also rained down from interplanetary space.

It is difficult to estimate the proportions in which these various sources contributed to the Earth's water content. But no matter how they stack up, it is reasonable to suppose that similar sources prevailed among Earth's planetary neighbors, Venus and Mars, so that they too must have been watery worlds in primordial times.

In the case of Mars, topological features resembling shorelines and riverbeds testify to the previous existence of standing and running water. Most of it has since evaporated into space. Mars, being a less massive planet than Earth and Venus, does not have enough surface gravity to have been able to retain a dense atmosphere for long. Ever since the atmosphere of Mars thinned out, liquid water on its surface has been prone to quickly boiling away.

The case of Venus is less obvious. It retains an atmosphere 50 times denser than that of the Earth and exerts 90 times the air pressure at the surface. Its closer proximity to the Sun, slow spin rate (225 times slower than Earth), and Earth-like mass may be counted among the key initial factors that, in combination, made the development of such a profound atmosphere inevitable. Consequently, surface temperatures on Venus now hover near 900°F. Once its atmospheric conditions began to resemble a high-pressure furnace, it stopped raining on Venus forever. Venusian heat would have elevated water vapor to such high altitudes that it was atomized into hydrogen and oxygen by incoming ultraviolet radiation from the Sun. The lightweight hydrogen atoms could then easily escape into space, leaving the heavier oxygen atoms behind to become integrated into carbon dioxide gas and other compounds. Water on Venus thus became even scarcer than water on Mars.

Clearly then, abundant water can only exist on other worlds under an auspicious confluence of favorable conditions. We commemorate the origins of that precious liquid, which makes life possible in this depiction of a volcanically active "water world" orbiting a star inside Ptolemy's Cluster. The planet is imaginary, but the star cluster is real. The 220-million-year-old cluster is so named because the first known mention of it was made by the famous ancient Greek astronomer. It is visible from Earth, almost 900 light years away, in the constellation Scorpius—five degrees north of the "Scorpion's stinger" marked by the stars Shaula and Lesath. From our vantage point within, the cluster stars gleam brightly in the twilight of an alien sky, which also features two nearby moons and a spectacular comet.

Next, let us explore the borders of the sea—land ho!

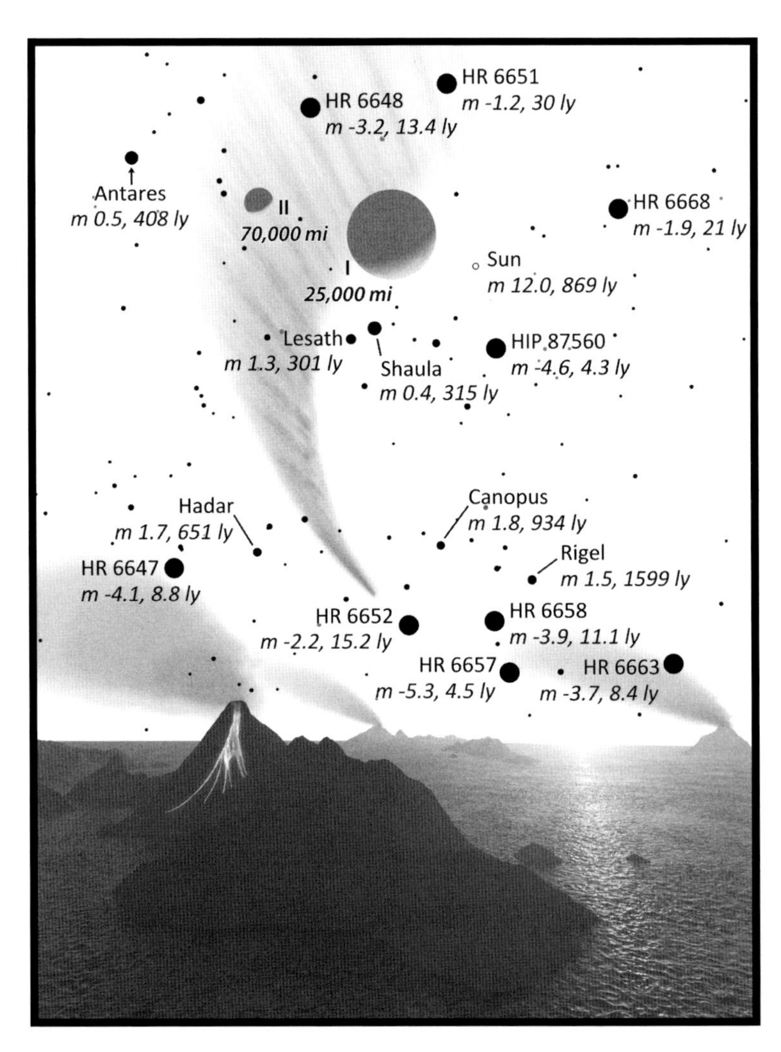

The Rise of Continents

Hyades Cluster

However vast and essential the oceans may be for an Earth-like planet, it is the intervening realms of dry land that are of central concern to humankind. The outlines of the Earth's continents are indeed the most iconic hallmarks of our world.

The rise of continents is made possible by plate tectonics. The Earth's crust is not monolithic. Rather, it is divided into enormous tile-like segments—*plates*—that are separated by various types of boundary zones that allow movement. The plates float like icebergs on a submerged ocean of churning magma—the Earth's upper mantle. The underlying turbulence convey the continents across great distances over geologic time scales—*continental drift.* Hence, the present-day configuration of continents on Earth is only temporary. A few hundred million years ago, all the land mass on Earth was fused together, forming a contiguous *supercontinent,* which geologists have dubbed *Pangea.* A few hundred million years hence, all the continents will rejoin into a new supercontinent. Cycles of dispersal and rejoining have been occurring on Earth since the very first continents arose, perhaps as early as four billion years ago. The first continents were modest in size. But over time, the total extent of land mass on Earth has tended to expand.

When we look to the famous Hyades Cluster, we find star systems that are about 625 million years old—the same age that the Earth was when it sprouted its first continents. We have encountered the Hyades previously in our excursions through space and time. In this scene, we visit a hypothetical Earth-like planet orbiting a Sun-like star within the cluster itself. As with the previous two star clusters we visited, the skies are crowded with exceptionally bright nearby stars. We may note the orange giant Aldebaran (#7 in the chart) lying toward the direction of the Sun (#4). For stargazers on Earth, the star *appears* to be the brightest star in the Hyades. But in this reverse perspective, it is rather subdued. Aldebaran is not actually a member of the cluster. Rather, it just happens to lie between the Hyades and the Sun (and closer to the latter!) at the present time.

Looking upon the landscape below, we find ourselves gazing across a deep chasm, filled in by a lake, that runs along a major geological fault line. Snow-capped peaks illuminated by low-slung moonlight indicate a cold climate at a high-latitude location. The sky's orange tinge betrays a rich presence of methane in the atmosphere.

A terrestrial planet without plate tectonics would not only be devoid of continents, but may also lack a *magnetosphere*—a planetary magnetic field such as the one that protects the Earth and its atmosphere from the high energy blasts of the solar wind. The generator of Earth's magnetism is believed to originate from convection currents in a liquid layer of Earth's core. This dynamo's power is determined by how well heat can escape the system, and plate tectonics may play a crucial role for efficient heat exhaustion.

Why tectonic activity is absent on Venus and Mars is a more difficult enigma to solve than the previous problem we considered regarding the absence of oceans. But the two problems may be interconnected. A dry planetary crust would be more rigid than one infused with the solvent of water. Hence, the presence of water may help to keep tectonic plates pliable and prevent the whole process from grinding to a halt.

On the other hand, we shall see that it would not be beneficial to have too much of a good thing.

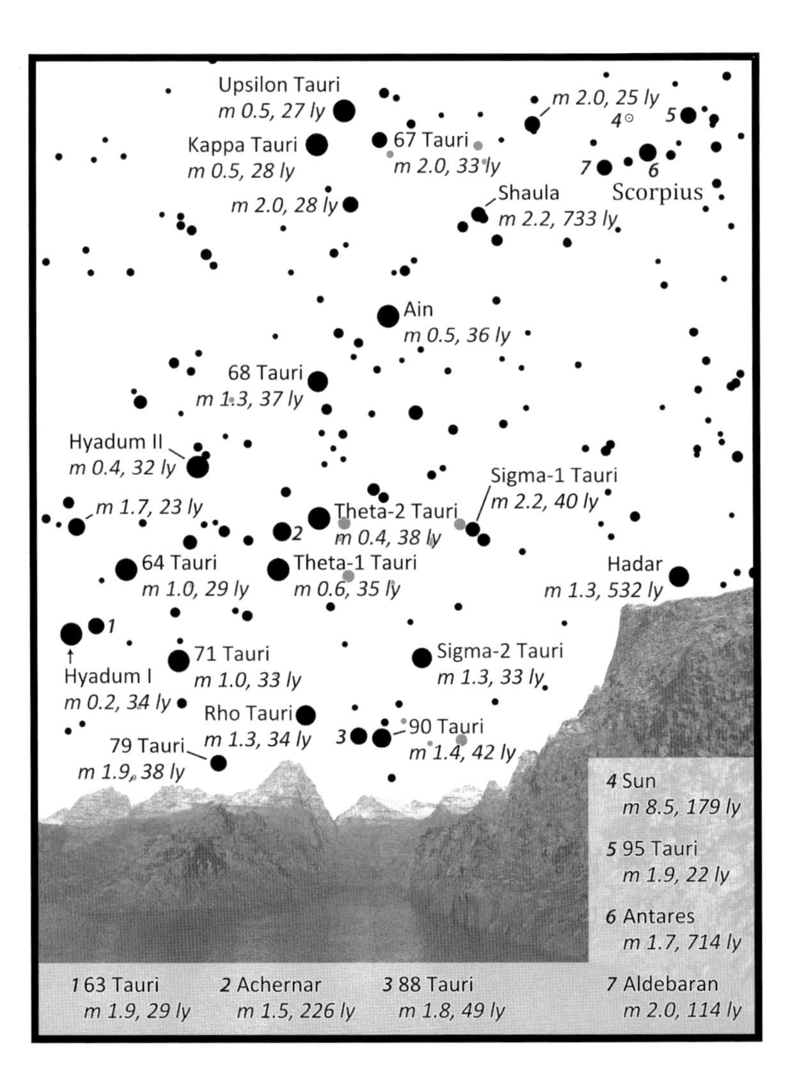

Leaving the Nest with Big Brother in Tow

Beehive Cluster (M44)

At the beginning of this itinerary, we witnessed the birth of a star in a molecular cloud complex. Such nebulosities typify the birthplaces of stars. But stars are almost never born alone. When the Sun ignited 4.6 billion years ago it may have been just one spark in a brood of hundreds or thousands of stellar siblings. For the many millions of years that these young stars stayed together, they formed a star cluster—similar to the ones we have just visited in our three most recent destinations.

As the stars grew older they escaped one by one, until their birth cluster completely evaporated. Where the sisters of the Sun reside today is anybody's guess. They would likely have a broad range of masses; hence, most of them would not have much physical resemblance to the Sun. By now, the Sun's most massive siblings have already died off and the least massive ones (which would have been among the first to leave the "nest") have already scattered to every sector of the Galaxy.

Life on Earth would have been in perpetual jeopardy if the Sun lingered in its birth cluster for too long. Its orbit around the cluster's central mass would have brought frequent close encounters with its co-orbiting siblings. A very-near collision would suffice to shuffle the orbits of its developing planets, or even cast them off into deep space. Moreover, massive stars are also prone to going nova—or even supernova. If massive stars were prevalent in the Sun's birth cluster, it would have been detrimental for the Earth to be subjected to nearby explosions on an ongoing basis.

In the scene presented here, we visit yet one more star cluster. The Beehive Cluster, also designated *M44*, is an open cluster seen in the zodiacal constellation Cancer the Crab from the skies of Earth. It is nearly the same age as the Hyades—perhaps slightly older. Once again, we find a Sun-like star that is offset by a fair distance from the cluster center. The time has come for the star to be released. It is leaving its siblings behind and will strike out its own path through the Galaxy.

We take in the view from five AU away from the star—behind an orbiting gas giant. We capture the moment when a large comet has been caught in the strong gravitational field of the Jovian world. The original nucleus of the comet consequently disintegrated into several fragments, each of which is sporting a separate tail blown back by the stellar wind. The leading fragments have already struck the planet, blackening the cloud tops at the sites of impact.

Some astronomers believe that Jovian worlds might function as "cosmic vacuum cleaners" of inner star systems. In our own solar system, comets impacting Jupiter in a fashion resembling the scenario depicted here have occurred as recently as 1994 and 2009. More importantly, Jupiter is thought to have cleaned up the bulk of stray comets in the inner solar system within the first several hundred million years of its history. Without Jupiter, the Earth might have been regularly bombarded by comets throughout its history and even today.

It is fortunate that the rain of comets, which helped to fill our oceans, eventually came to a stop; thus the way was cleared for life.

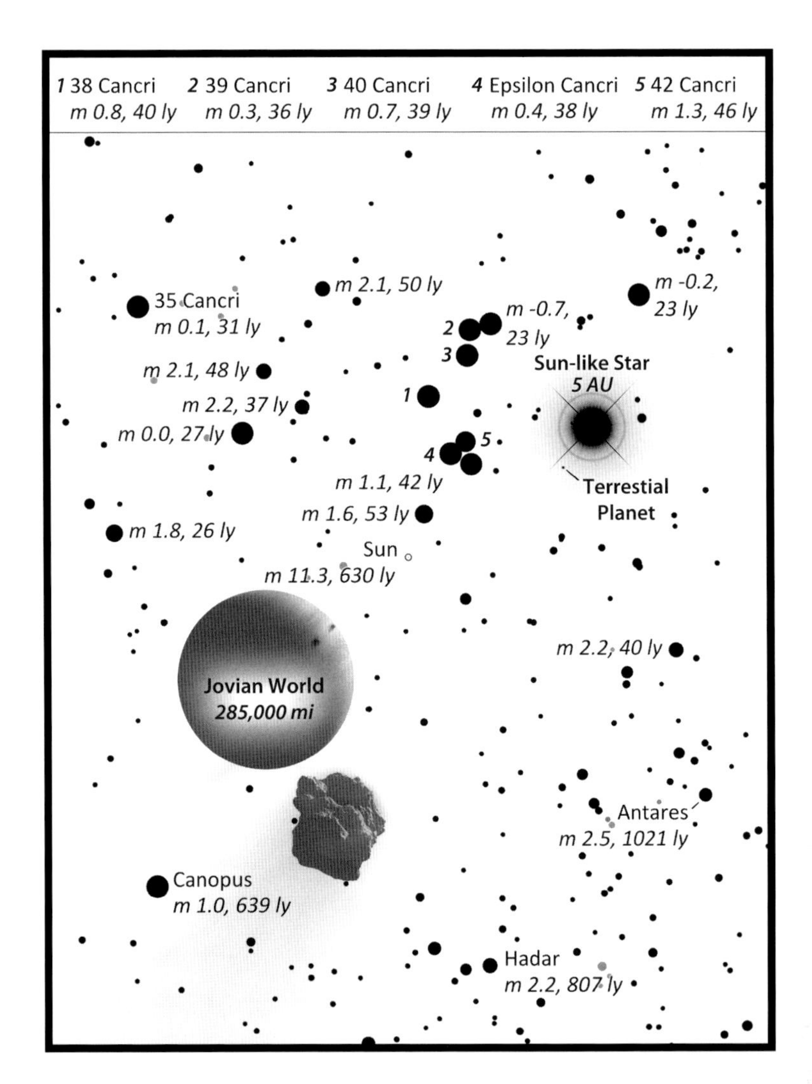

Oxygen Revolution

18 Scorpii

Earth-like planets, to be worthy of the definition, ought to be life-bearing planets. How life on Earth first arose is difficult to discern, but the timing of when life first arrived can be traced back to when the Earth was roughly one billion years old. The evidence is preserved in the form of stubby sedimentary pillars (discovered in Australia) that resemble structures formed by colonies of contemporary microorganisms. These ancient *stromatolites* are our earliest line of indirect evidence. By the time the Earth was two billion years old, direct evidence of microorganisms is attested by microscopic fossils and chemical signatures of life.

Chemical evidence also indicates that a key change in the Earth's environment occurred when the Earth was 2.2 billion years old. Photosynthesizing bacteria (which convert sunlight into food by the same process that plant life does today) eventually saturated the Earth's atmosphere and ocean shallows with their primary waste product: oxygen. The sudden oxygenation had monumental consequences for life on Earth. Most of the planet's inhabitants living before then were *anaerobic* bacteria—organisms that did not require oxygen to live. Even worse, oxygen was actually harmful to many of them. A mass-extinction among anaerobic life-forms was the likely outcome. In the wake of this "oxygen revolution" however, new opportunities were created for life to explore. The biological domain of the *Eukaryotes* arose in the aftermath—cellular life forms that are vastly more complex than bacteria. These types of organisms would eventually seize upon *aerobic* (oxygen-burning) metabolisms and blaze evolutionary pathways to diverse forms of energetic life.

We now visit another hypothetical Earth-like world, which may parallel the conditions on Earth when the oxygen revolution began. The planet is located in the environs of the Sun-like star 18 Scorpii. Ascertaining the age of individual stars is much dicier than doing so for star clusters. An age of 2.4 billion years has recently been assigned to this star, but we can only be certain that its age does not exceed 6.8 billion years. Nearly the Sun's equal in luminosity and metal content, 18 Scorpii is located 45 light years away from Earth, where our Sun appears as a star of magnitude 5.5 under the "hooves" of Taurus the Bull.

We behold a moonlit view of an alien seashore. The incoming waves lapping upon the dark sands of the beach are laden with sea foam—a sign that these waters contain organic matter. A short distance offshore stands a craggy outcrop in the form of a natural bridge, allowing a peek to the distant horizon where the bright star Rigel can be sighted.

Another intriguing postscript to Earth's oxygen revolution was that an ice age occurred in its wake—when our planet was 2.3 billion years old. The spell was severe. Earth's oceans may have *completely* frozen over once or twice in this period—a condition that geologists have dubbed *Snowball Earth.* The timing of this climatic upheaval suggests that it was triggered by the oxygenation event itself. If so, we must wonder why the ice age subsided even though atmospheric oxygen persistently increased.

Indeed, much later in Earth's history (850-635 million years ago), yet another severe ice age occurred, bringing more Snowball Earth episodes. This latter period was followed by another dramatic revolution for life on Earth, which also may be paralleled today on other worlds.

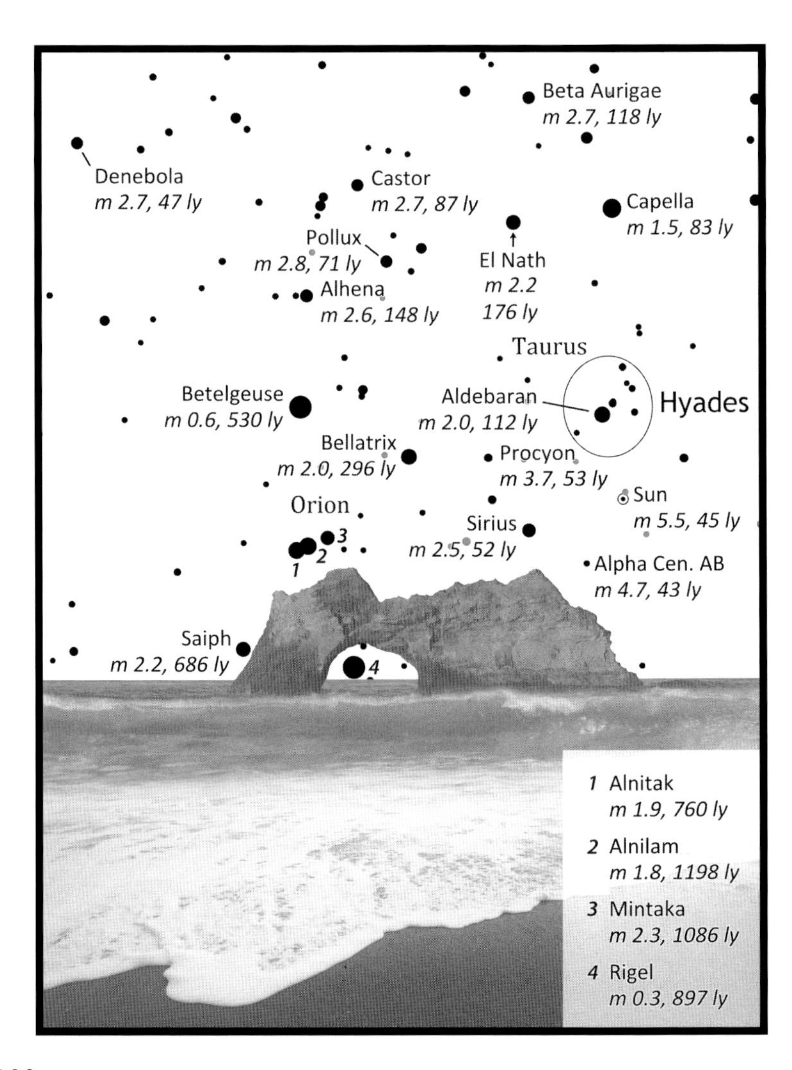

Evolutionary Explosion

Zeta Reticuli

The evolution of life on Earth was a process that advanced at an irregular pace. Long periods of stagnation were sometimes interrupted by sudden fits of innovation. The most extraordinary period of rapid evolution in Earth's history was the *Cambrian Explosion*—an interval of several million years that transpired when the Earth was 4 billion years old. Ocean life grew explosively in size and diversity, which is attested to by vast collections of treasured fossils dating back to that time.

Why did the Cambrian Explosion occur when it did, after microscopic life had steadily ruled the Earth for billions of years? The puzzle is difficult to solve. The advancement may have been prodded by favorable changes in the environment, or by the crossing of some critical threshold of evolutionary development, or by sheer fits of spontaneity, or by some fortuitous combination of all of the above. In any case, the end-results were spectacular. The proliferation of complex multicellular life at that time was an essential step toward producing the most conspicuous species of animal life in our world today—including human life.

Could a similar stage of rapid evolution be occurring elsewhere in the Galaxy today? In the scene depicted here, we imagine it happening on a hypothetical Earth-like world in the Zeta Reticuli star system.

Zeta Reticuli is a widely-separated binary system comprised of two Sun-like stars. Even at 39 light years away, the duplicity of the system can be discerned by the naked eye in the southern hemisphere skies of Earth, where it appears as a close pair of fifth magnitude stars. The angular separation is one-twelfth of a degree (one-sixth the diameter of the full Moon). The spatial separation of the two component stars—designated $Zeta^1$ and $Zeta^2$ Reticuli—is approximately 4,000 AU. The pair's orbital period around their common center of gravity is unknown.

The Galactic orbit of the system is moderately eccentric (35%); its distance from the Galactic center varies from 13,000 to 27,000 light years. The age of the system is likely to be 2 or 3 billion years old, but certainly no older than 7 billion years. The pair is somewhat less abundant in heavy elements than the Sun (60% of the solar value). $Zeta^2$ Reticuli shows signs of circumstellar debris disks that are similar in size to the asteroid belt and Kuiper belt in our own solar system.

From the point of view of this hypothetical world depicted to orbit $Zeta^2$ Reticuli, both stars are readily visible in broad daylight. $Zeta^2$ appears as bright as the Sun from 1 AU away, while the much more distant $Zeta^1$ appears at magnitude -8.6 (30 times brighter than Venus at its brightest). The light streaming in from $Zeta^2$ Reticuli reveals an apparently desolate seascape. But in producing this portrait, our photographer attached a polarizing filter to his camera lens, which helps us penetrate the sea surface reflections. This gives us a glimpse into an underwater arena, where we find exotic forms of life therein. A few of these resemble corals. Another group resembles jellyfish. Another individual swimming nearby resembles the kinds of segmented invertebrate animals that perhaps typified the Cambrian Explosion.

Complex life may have conquered the seas here, but the realm of land remains barren. Perhaps more courageous steps upward are being taken elsewhere.

Invasion of the Land Dwellers

HD 197210

Following the Cambrian Explosion, another major transition in the evolution of multicellular life on Earth occurred—the transition from sea to land habitats. Algae gave rise to land vegetation, arthropods gave rise to insects, and fish gave rise to amphibians. The transition of fins into limbs is perhaps the most iconic evolutionary transformation in all of natural history. All land vertebrates that roam the Earth today descended from a common ancestor that crawled out of the sea a few hundred million years ago.

What lured complex life to abandon its watery domain? Perhaps it was a less competitive environment for untapped resources (nutrients, oxygen, and other sustenance) that may have been one enticement. A refuge from sea-bound predators may have been another. Perhaps we may even dare to suppose that primordial amphibians simply had a taste for adventure. What, after all, stirs the modern human desire to attempt space travel? It may stem from a genetic inheritance that is very ancient indeed.

Let us now imagine a replay of the ancient drama of an evolutionary land invasion—on a hypothetical Earth-like planet 100 light years away from home. HD 197210 is a Sun-like star in the constellation Aquarius. It is estimated to be 4.4 billion years old but could possibly be much older. The star is 83 percent as luminous as the Sun and its abundance-level of heavy elements is 95 percent that of the Sun. Its Galactic orbit is moderately eccentric—similar to that of the Sun and typical of many other stars in the Sun's age range. The orbit of HD 197210 carries it between 20,000 and 27,000 light years from the Galactic center.

The scene depicted here portrays an Earth-like world orbiting HD 197210. A bright moon rising behind us illuminates a cloudless evening. We behold a muddy shore at the edge of a large bay. Primitive forms of terrestrial plant life with curled stems on upright stalks have taken hold at the water's edge with horizontal roots. Crawling forth from the brackish shallows are two fish-like amphibians; they breathe the air with their primordial lungs while taking in the view.

High above this budding landscape we once again find the Hyades star cluster in the top right corner of the portrait. It is bordered by brighter stars that Earthbound stargazers find in Taurus and Orion. The Sun-like star 11 Aquarii also shines brightly nearby from just 13.4 light years away. We may also find the familiar stars Capella, Vega, Altair, and Sirius on the chart. The position of the Sun lies midway between the latter two, but too dim to be visible in this portrait at magnitude 7.3.

As for our amphibian invaders, they seem disinclined to venture much further into the immediate frontier—at least for now. The rocky hillside in front of them appears to be inhabited only with green patches of moss. A geyser in the distance blows a jet of hot steam into the starry sky. Perhaps in due time these pioneering invaders may spawn future generations of varied descendants that participate in a more highly developed ecosystem rivaling the complexity of life on Earth.

Perhaps elsewhere in the Galaxy, this has happened already.

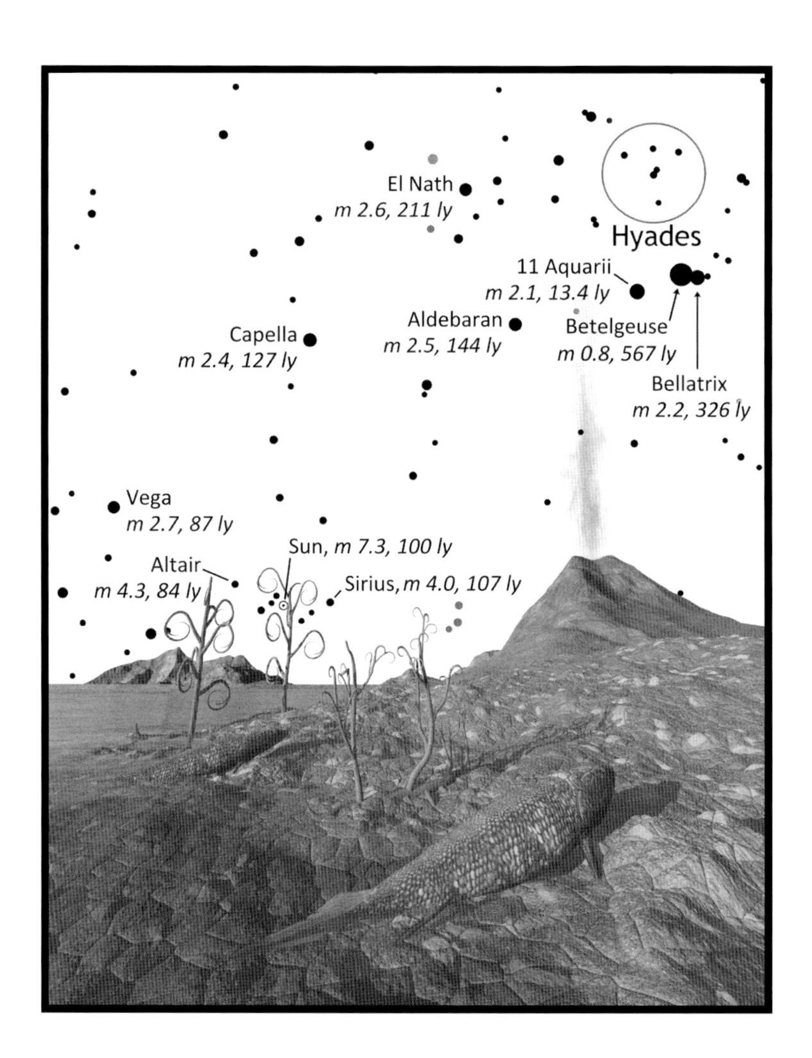

The Synthesis

Alpha Mensae

Our last itinerary is nearly at a close. So far, we have witnessed the formation of planets, moons, oceans, continents, biochemistry, and advanced life forms. We have recognized each of these features as key stages of development for potentially Earth-like worlds. But we may not have even nearly covered everything. Our Earth may enjoy other special circumstances that are more difficult to recognize—or some even so subtle that we completely take them for granted. But now that we have covered these essential attributes, let us visit one last hypothetical world that synthesizes all of the elements we have noted.

Alpha Mensae is a fifth magnitude star in the skies of Earth—the brightest star in the faint modern southern constellation Mensa the Table. Located 33 light years away from the solar system, it is orbited by a recently-discovered red dwarf companion, B, at an approximate distance of 30 AU (equivalent to the distance between the Sun and Neptune). The main star, A, is 93 percent as luminous as the Sun and slightly more abundant with heavy elements. It is estimated to be 5.8 billion years old, but it almost certainly cannot be any older than 11 billion years. That system's Galactic orbit is somewhat more eccentric than that of the Sun. It is carried around the Galaxy at a distance from the Galactic center that varies between 17,000 and 26,000 light years.

No planets have actually been discovered in the Alpha Mensae system. But in the scene depicted here, we imagine the star to host at least two hypothetical worlds. Arrayed immediately before us is an Earth-like terrestrial world. We behold a coastal landscape at sunset. The environment is populated with grasses and tall trees, with needled leaves, that resemble spruces. A flying reptilian-like creature executes a swooping banked turn over our heads while en route to its nest. In its clutches it carries dinner to its family—a freshly caught fish plucked from the ocean below.

Above the scattered clouds that fill the evening sky, we catch sight of Alpha Mensae B. The red dwarf star is plainly visible in broad daylight at magnitude -11, which is equivalent to one-fifth the brightness of a full Moon on Earth. In the upper right of the portrait, we also catch a glimpse of a hypothetical gas giant similar to Jupiter. This jovian world, a swallower of comets, is also a protector of life.

What might this world's future potential be? Are we witnessing an era that parallels the great age of dinosaurs on Earth? Might the creatures of Alpha Mensae eventually fall into extinction, as the dinosaurs did on our world tens of millions of years ago, or could they stably endure for all the eons that this world will continue to exist?

Further evolutionary advancement is surely possible. We recognize that it happened on our own world. We even explored a tiny fraction of it on our second itinerary, when we followed the course of Spaceship Earth during the rise of apes, people, and civilization.

Do other civilizations exist elsewhere in the Galaxy? This is usually what the human imagination has in mind when pondering the idea of *extraterrestrial life.* We noted in our first itinerary the failed first attempts by radio astronomers to find radio signals broadcast by advanced civilizations from nearby stars (page 14). Renewed efforts are ongoing. Let us now return to Earth and tune in for an update.

Are We Alone?

The Search for Extraterrestrial Intelligence

The discovery of intelligent life beyond Earth would be a cultural event rivaling no other in human history. With the advent of the radio telescope in the 20th century, it was hoped that the first interception of interstellar communications would be imminent. But in the five decades since, not a single message from the stars has ever been received.

It is not for lack of trying. Despite the withdrawal of support by government-funded agencies since the 1990's, the *Search for Extraterrestrial Intelligence* (SETI) has persisted through private means. Project Phoenix, for instance, resurrected from the ashes of a cancelled NASA program, surveyed 800 stars within 200 light years of Earth using large radio telescopes all over the world. Billions of frequency channels were sampled for each star. It took nine years to work through the list of targets. But even this much thoroughness could not churn up anything beyond the monotonous crackle of radio static.

Why the silence? Are we alone? Could human society be the only advanced civilization within 200 light years of Earth? Or even within our whole Galaxy? Questions like these inevitably conjure up two long-debated topics: the *Fermi Paradox* and the *Drake Equation*.

The Fermi Paradox originates from a legendary conversation between physicist Enrico Fermi and a few of his colleagues at Los Alamos National Laboratory circa 1950. En route to lunch, they happened to be chatting whimsically about flying saucers and the feasibility of interstellar space travel. The conversation moved on to other topics but Fermi, who was renowned for making accurate estimates from minimal information, brooded. Some time after being seated at the lunch table, he suddenly blurted out, "But where is everybody?!" The outburst provoked laughter, but his implication was serious: if our Galaxy is teeming with space-faring civilizations, then many of them ought to have reached Earth by now. But because they have not (it was plainly obvious to the Los Alamos physicists that reported sightings of flying saucers were false), then perhaps we must concede that civilized worlds cannot be prevalent.

One of the other participants in the conversation, Edward Teller, recalled much later: "I do not believe that much came of this conversation, except perhaps a statement that the distances to the next location of living beings may be very great and that, indeed, as far as our Galaxy is concerned, we are living somewhere in the sticks, far removed from the metropolitan area of the Galactic center." Does this explain SETI's failures? Could civilized worlds be spread out so sparsely in our region of the Milky Way that even radio communication between them becomes impractical?

This sort of "population density" problem leads us to the other related topic: the Drake Equation. Frank Drake, we may recall, was the pioneering radio astronomer who led Project Ozma at the Green Bank Observatory in 1960 (page 14). The following year Drake convened a meeting of scientists and technicians at the observatory on the topic of SETI. As points for discussion, he outlined the prospects for detecting radio signals from other worlds as a series of probabilities applied to questions such as: What fraction of stars in our Galaxy have planets? What fraction of planets can support life? What fraction of planets with life could eventually develop intelligent life? By making educated guesses for each value, one may lay odds on the ultimate prospects for SETI's success.

But because most of the values in the Drake Equation are purely speculative, the resulting estimates can vary widely from one person to the next. Several variations on Drake's original formulation are also in vogue. Carl Sagan, for instance, worked through a version of it in his 1980 hit television series *Cosmos*. Beginning with the premise of 400 billion stars in the Galaxy hosting one trillion planets, he went on to surmise that life may have evolved on 100 billion worlds, and that advanced civilizations may have emerged on one billion worlds. He then ascribed the most abysmal fractional value to the number of civilizations that may have survived to the present day—one out of 100 million—on the premise that any technological civilization may only be expected to last for several decades before it self-destructs in a fit of global nuclear warfare. This decimates the number of surviving civilizations to a mere *ten* in our entire Galaxy—a chilling reminder of the Cold War era pessimism that overshadowed Sagan's show. But he concluded his dramatization with an alternative value: maybe as many as one out of a hundred may survive, leaving us 10 million Galactic civilizations to search for.

One may critique Drake's approach on many levels. Many critics have. All we can say for sure is that the population density of the Galaxy can only be settled by *observation*—not argumentation. Meanwhile, SETI stills draws blanks. What else can be done?

Space Scopes and Starshades

Pursuing Pale Blue Dots

In 1990 NASA launched the Hubble Space Telescope, the world's first general-purpose space observatory. The portraits of the universe that the Hubble has sent down to us rank among the most inspiring and widely-distributed images ever produced by science.

NASA's next-generation space observatory, the James Webb Space Telescope (JWST), will surpass the Hubble's capabilities. Its primary component is a curved mirror comprised of hexagonal segments made of polished beryllium. They are plated with gold to enhance the mirror's reflectivity in the spectral range of infrared light. Its total surface area is seven times larger than the Hubble's primary mirror. The optics are not enclosed in a large tube. Instead, a multi-layered *sunshield* at the base of the telescope will prevent any interference of heat and light from the Sun, Earth, and Moon.

Once it launches (sometime in the year 2018 or afterward), JWST will be placed at *L2*—a region of gravitational stability in the solar system located one million miles from Earth, directly opposite of the Sun. JWST will be capable of photographing galaxies at the very edge of the visible universe—penetrating new depths of the cosmos that have so far remained unseen by any previously existing telescope.

But just like the Hubble before it, JWST may also be configured for a wide variety of research. One useful accessory proposed for JWST is a *starshade*—an opaque daisy-petaled disk some 100 feet in diameter that could be launched subsequently and rendezvous with JWST at L2. The starshade would simply float in space a few tens of thousands of miles in front of the telescope itself. It would be actively maneuvered into whatever direction that the telescope is pointing. Any star centered in the telescope's view would be concealed by the starshade, while the star's immediate surroundings remain unobstructed. A star system observed in this fashion makes it possible to photograph exoplanets *directly*.

Why is such an elaborate system necessary for faraway exoplanets to show up in a photographic image? The situation is analogous to a firefly standing on a distant searchlight beaming toward us. It would be impossible to discern the firefly in the searchlight's overpowering glare. Place a lid in front the searchlight, however, and the firefly will pop into view. This is essentially how a starshade functions: it shutters starlight so that fainter planets nearby can be seen.

The portrait on the opposite page shows the fully deployed James Webb Space Telescope. The inset at the top right simulates how a star system might appear when observed with a starshade in place. The starshade blacks out the alien sun, outlined by the glow of dust and debris in the star system. And the pale blue dot off to the side? That's what an Earth-like world would look like. The image is unspectacular—not the alluring vision of "oceans, clouds, continents, and mountain ranges" that we contemplated at the outset of this itinerary. But it would be enough to make great discoveries. One can learn a lot from a dot. By monitoring any changes in its brightness and color, one might be able to tell if the planet is rotating. Longer term variations might indicate seasonal icecaps or perhaps even seasonal vegetation. A spectral analysis could reveal the chemical composition of the planet's atmosphere—perhaps even the presence of *biomarkers* (signs of life) such as free oxygen (page 102).

Star systems within 30 light years of the Earth would be the easiest targets for JWST to image. These include all the stars that we visited in our first itinerary, and an abundance of nearby red dwarf stars like Gliese 876 and Gliese 581. We had asked if planets can be "habitable" in red dwarf systems (page 78). We may soon find out by observing planets in red dwarf systems directly.

Here is a more interesting question: how many planets within a volume of space nearest the solar system will be found showing *telltale signs of living ecosystems?* A single example within 30 light years would be awe-inspiring; multiple examples would be mind-blowing. The tally would give us a clue about the Galactic density of living worlds. Are there pathways for interstellar migration open to future generations of human space travelers—or to alien space travelers already? If so, Fermi's Paradox suddenly becomes more disconcerting than ever: *where are they?* Maybe we *are* alone.

On the other hand, maybe we won't find any pale blue dots in the Earth's immediate vicinity. Maybe we will just find the pale red dots of desert worlds, the shiny white dots of snowball worlds, or no dots of the terrestrial kind at all. We might then find ourselves "living somewhere in the sticks," just as Teller suggested. We would then *still* be alone—maybe not the sole civilized world in the Galaxy, but isolated all the same.

Let us have the courage to keep exploring, no matter what we may find.

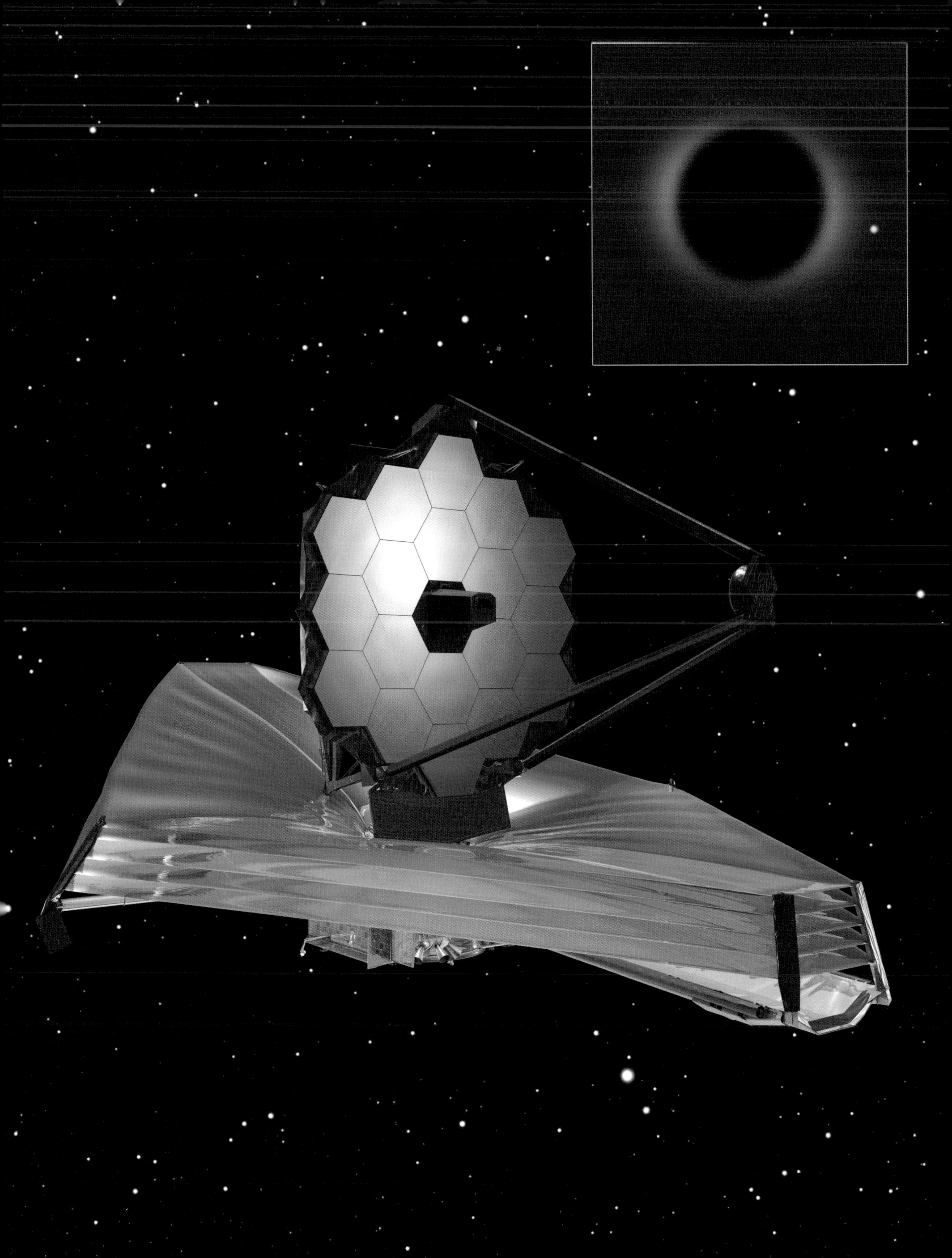

EPILOGUE

The Future of Galactic Exploration

The Drive to Discover

Sharp Turns Ahead at the Edge of the Unknown

"The definitive study of the herd instincts of astronomers has yet to be written, but there are times when we resemble nothing so much as a herd of antelope, heads down in tight formation, thundering with firm determination in a particular direction across the plain. At a given signal from the leader we whirl about, and, with equally firm determination, thunder off in a quite different direction, still in tight parallel formation."

- J. Donald Fernie, Astronomer, 1969

In recent times, the growth of knowledge afforded by the sciences has been tremendous. Advances in modern astronomy, made possible by theoretical physics and telescopic observation, have allowed us to survey and comprehend the universe in robust detail—from the subatomic processes that drive the lives of stars to the structural complexities of galaxy clusters on a cosmic scale. We can rightly take pride in these accomplishments. Indeed, the history of science is often written up as a series of spectacular demonstrations of human ingenuity.

However, the inclination of historians to direct attention toward scientific successes also has the potential to mislead us. It fills us with expectations that the current research programs garnering the most attention in the news today are also poised for imminent successes. A critical review of history makes us wiser. History shows that scientific ideas that eventually triumph are often eclipsed by unsuccessful ideas for extended periods of time before the former are allowed to come to light.

Astronomy abounds with such examples, beginning with the ancient Greeks. The Earth-centered model of the solar system perfected by Ptolemy in the 2nd century CE prevailed for many centuries, while the first Sun-centered model, proposed four hundred years earlier by Aristarchus, languished in obscurity. After Copernicus reprised the idea in 1543 (the year of his death), several more decades elapsed before his work was widely recognized as revolutionary. This occurred when the Copernican model was championed by Galileo, who was infamously punished for doing so.

Another controversy that took centuries to resolve was the nature of our Galaxy and whether or not it was the only one in the universe. In the wake of the Copernican revolution (and despite Kepler's stern disapproval, noted on page 50), the stars in our skies began to be recognized as distant suns, strewn through unlimited space.

This picture of the universe inspired Isaac Newton, who devised the principle of universal gravitation in 1687, to insist that the universe was infinite—otherwise, he thought, all the stars in the universe would be drawn to the cosmic center by their mutual gravity and collapse into a colossal cosmic heap. Because an infinite universe has no conceivable center, Newton thought the forces of attraction between stars could thus remain eternally balanced.

In 1755, the German philosopher Immanuel Kant devised another solution. Just as the orbital motions of planets in our solar system keep them from falling into the Sun, Kant proposed that the stars too were in perpetual motion, orbiting the center of gravity of a common system. This system was deemed to be an "island universe" existing in relative isolation from the rest of the cosmos. In the same treatise, Kant suggested that the cloudy apparitions of lenticular (elliptically shaped) nebulae observed by astronomers in his day might be other island universes—too distant to be resolved into individual stars.

In the following century, larger and more powerful telescopes revealed that many such nebulae exhibit spiral patterns. They were thence dubbed *spiral nebulae.* But because it was not yet possible to discern their distances, it was likewise impossible to tell whether the spiral nebulae were indeed island universes, as Kant originally proposed, or small nebulous disks inside our own Galaxy.

The consensus was firmly resolved on the latter when the astronomical historian Agnes Clerke wrote in 1905, "The question whether nebulae are external galaxies hardly any longer needs discussion. It has been answered by the progress of research. No competent thinker with the whole of the available evidence before him can now, it is safe to say, maintain any single nebula to be a star system of coordinate rank with the Milky Way."

Yet, twenty years later, the consensus reversed itself. The famed astronomer Edwin Hubble firmly established that the two nearest spiral nebulae were at least a million light years distant. These objects, and billions more, were indeed island universes—galaxies akin to the Milky Way.

What is the moral of these cautionary tales? It is that history is full of surprises. All that we can be sure of is that we will be surprised again.

The Gaia Mission

Mapping the Galaxy with Precision

Late one afternoon in 2013, a Soyuz-Fregat rocket sits on a launchpad at the Kourou spaceport in French Guiana. Its payload: a two-ton spacecraft named *Gaia* built by the European Space Agency. When Gaia launches, it will be sent to *L2* (see page 112). Its five year mission is to succeed Hipparcos (page 8) in mapping the stars.

Hipparcos' successor has been long overdue. In the two decades following the Hipparcos mission, no new space-based astrometry mission has ever been launched. Several missions proposed in the interim have failed to muster budgetary approval. Prospects for major advancements in Galactic astronomy now rely heavily upon the Gaia mission. If fully successful, Gaia will catalog the positions, brightnesses, colors, and motions of a billion stars in our Galaxy that are brighter than magnitude 20. Gaia's astrometric precision will range between 50 and 200 times finer than that of Hipparcos. Hence, whereas Hipparcos could usefully measure the parallaxes of stars as far away as 1,000 light years, Gaia will fix distances of stars as far away as the Galactic center with comparable accuracy, and nearby stars with incredible accuracy.

Plotting Gaia's inventory of stars in three-dimensional space will give us a Galactic map broad enough to exhibit large swaths of the Galaxy's spiral arms and other large-scale features. This will resolve any lingering confusion about the basic structure of the Galaxy, such as how many spiral arms the Galaxy has, how they are configured, where exactly the Sun resides, and the size and orientation of the Galaxy's central bar.

The proper motions of Gaia's target stars will also be measured with unprecedented precision. A by-product of these measurements will be the discovery of many new exoplanets and the refinement of mass estimates for many exoplanets already discovered. Incidental detections of transiting exoplanets (the same technique used in the Kepler mission, page 82) will also be inevitable.

Unlike Hipparcos, Gaia will be equipped with a spectrometer capable of producing radial velocity measurements. This will allow the motions of the one billion stars in Gaia's catalog to be projected into three-dimensional space. The spectrographs recorded by Gaia will also indicate the abundances of various elements for every star. Astrophysical models of stellar evolution, which often yield unreliable ages for individual stars, will likely be improved by utilizing Gaia's immense database of spectra and precise parallaxes.

These are the main mission objectives that Gaia's designers expect to fulfill. But the most valuable aspect of Gaia's database is that it will allow researchers to explore the Galaxy in original, unanticipated ways. A general rule of thumb in experimental science is that whenever measurement precision improves by an order of magnitude (ten times), major discoveries follow. When the 24-year lapse in precision star-mapping is remedied by Gaia, astrometric precision will be improved by *two* orders of magnitude (one hundred times). History-making surprises are surely in store.

Gaia even has the potential to open up new vistas in fundamental physics. One of the biggest mysteries in contemporary astronomy is the so-called "galactic rotation problem."

For many decades it has been known that the radial velocities of stars measured in the outer regions of galaxies signify orbital speeds that are curiously fast. The larger the orbits of the stars, the more their speeds seem to exceed the ability of gravity to hold them. The predominant theory accounting for these discrepancies assumes that there is unseen matter exerting additional gravity—a *halo* of invisible and intangible particles (cold dark matter)—that surrounds every galaxy. A rival approach is to rewrite the laws of gravity so that the strength of gravity exerted by normal matter is boosted at long distances. Both ideas have conceptual problems.

Stars in small orbits (e.g., within several thousand light years of a massive spiral galaxy's center) do *not* exhibit extraordinary speeds. Does it follow that the innermost regions of galaxies are swept clean of cold dark matter? If so, then how? The absence of dark matter in the interior of our own Milky Way Galaxy may even extend beyond the radius of the Sun's orbit, according to a 2012 study of star motions conducted at ESO's La Silla Observatory. Another 2012 study (University of Bonn) argues that the configuration of Milky Way satellite galaxies, globular clusters, and star streams implies that cold dark matter may not even exist in the entire universe.

Meanwhile, those who propose to modify the laws of gravity precariously claim that the gravitational boost remains inert until it suddenly "switches on" at an arbitrary threshold. For this reason the theory notoriously fails at large scales (i.e., it cannot completely explain the motions of galaxies in some galaxy clusters).

Scrutinizing the motions of stars in our Galaxy with Gaia's precise measurements will shed new light on these ongoing controversies.

Gaia
arianespace

Interstellar Migration

Homesteading the Final Frontier

In the realm of science fiction, interstellar space travel is customarily portrayed as a routine affair. In the future, starships that voyage through the Galaxy are no more uncommon than the ships of sea that navigate the waters of Earth today. Sci-fi authors often furnish their protagonists with vehicles and communication devices capable of transcending lightspeed—a necessary plot device for any storyline that requires prompt interaction between exotic locations separated by light years of space.

Designers of actual starships, however, will not have the luxury of circumventing the most well-established principles of physics. Transporting human passengers to the stars presents bewildering challenges. To propel a spacecraft even near the speed of light, and to decelerate it at a chosen destination, would require the generation of phenomenal levels of energy. It may not ever be feasible to build a starship that is capable of carrying human passengers to other star systems any sooner than the duration of a human lifetime.

Starships may then have to be artificial worlds unto themselves, colossal in scope, and capable of sustaining multiple generations of people for centuries at a time. Could human beings survive in comfort and sanity in such confinement throughout their entire lives? It would be convenient if the passengers could hibernate through their voyages in prolonged states of suspended animation. This too is a staple of science fiction, the feasibility of which remains unknown and unproven today.

No matter what futuristic technologies may become available to alleviate the difficulties, it is hard to imagine that interstellar space travel will ever become "routine." Every successful voyage to the stars will be an extraordinary historic event requiring enormous financial commitments and heroic personal sacrifice.

What destinations might lure the dauntless voyagers of posterity? If Earth-like planets truly capable of supporting human inhabitants can be found, some of our descendants will surely be inspired to colonize them. Others may wonder how such an undertaking could possibly be worth it. Why endure the costs, the risks, and the hardships of sending off a small band of colonists on a one-way trip to a primitive living environment?

On the other hand, fish had no reason to climb out of the sea, apes had no reason to abandon their trees, and Homo sapiens had no reason to fan out from East Africa. But yet, where would we be without them?

When the mountaineer George Mallory was asked why he wanted to climb Mount Everest, he famously quipped, "Because it's there." He and his climbing partner lost their lives on their journey, but we recognize that the spirit of their effort is an indomitable component of human nature.

Humankind will one day reach for the stars. Whether we are ready to embark in 500 years or 50,000 years is irrelevant. Any span of time in this range is a mere eye blink in relation to our long evolutionary odyssey. What vistas will our descendants behold?

In the unknowable future, on a planet unknown to our generation, a mother and her child stroll across a grassy plain in the twilight of evening. In the sky above, the Sun that hosts the homeworld of humanity is imperceptibly faint, but the child's mother indicates the location of Capella—a bright star nearby. "We came from a star beyond that one," she explains while telling the tale of their ancestors, who ventured beyond the solar system and into the Galaxy at large.

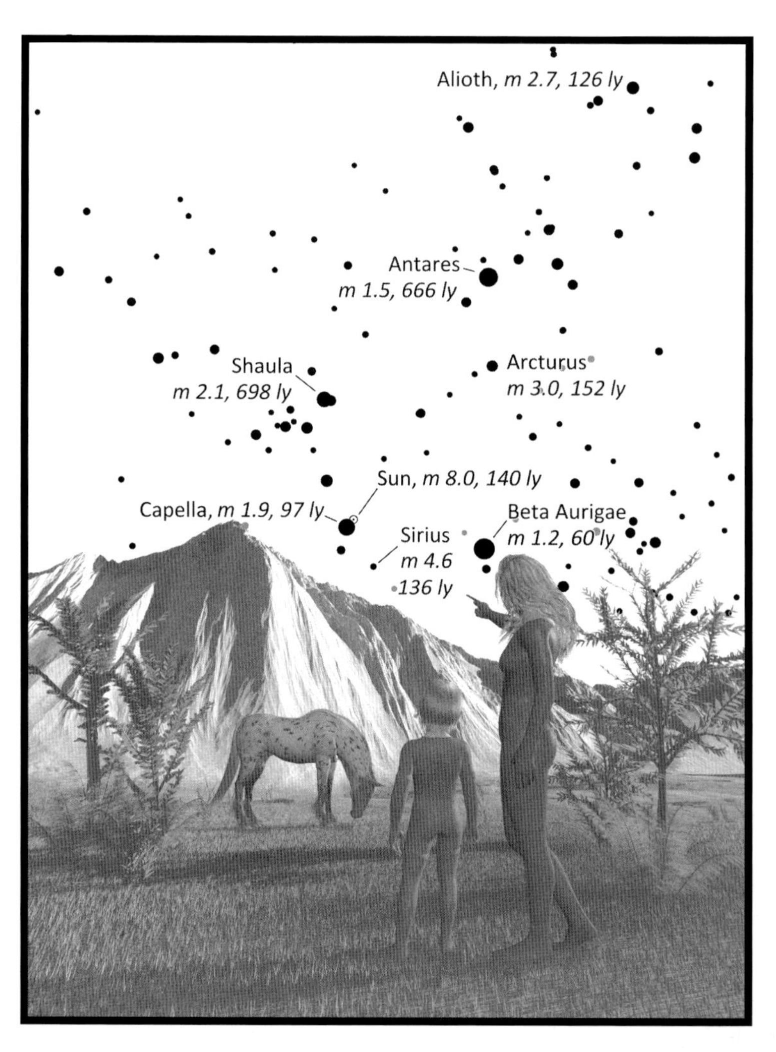

ACKNOWLEDGEMENTS

Data: Star positions, motions, and photometry based on *XHIP: an Extended Hipparcos Compilation*, a free database available through astrostudio.org. Exoplanet facts and figures are taken from The Extrasolar Planets Encyclopaedia at exoplanet.eu

Illustration tools: Starfields and charts were rendered in Inkscape (inkscape.org). Original planets, asteroids, terrain, and landscapes were created with Terragen 2 version 2.3 by Planetside Software (planetside.co.uk). Additional content was rendered in Daz Studio 4 (daz3d.com). General post-processing performed with Adobe Photoshop CS5 (adobe.com).

Resources: New World Digital Art (nwdanet.com); Marc Gebhart (web.me.com/marcgebhart1/Trees/ProTrees.html); Xfrog (xfrog.com); The Planetside Forums (forums.planetside.co.uk), with special thanks to Forum users *DandelO, HollyFlame, FrankB, Kevin F, cyphr, frelancah,* and *njeneb* for freely sharing example Terragen settings and techniques and to *Tangled-Universe* (Martin Huisman) for assistance with setting up NWDA products; Texture maps from JHT's Planetary Pixel Emporium (planetpixelemporium.com) and Tony Textures (tonytextures.com); Custom Photoshop brushes: Tree Brush Pack from Brusheezy (brusheezy.com) and Flames Brushes by Obsidian Dawn (obsidiandawn.com); Daz 3D models (daz3d.com); *Homo Erectus* by MEC4D (mec4d.com); *Wolf* by Alessandro Mastronardi (alessandromastronardi.com); Other 3D models from 3D Rivers (3drivers.com) and Falling Pixel (fallingpixel.com); Orbiter 2010 Space Flight Simulator (orbit.medphys.ucl.ac.uk); Stonehenge 3D Walkthrough (pixelparadox.com/stonehenge.htm).

Image credits: The image in the *Introduction* is based on a photo of the Getty Villa in Malibu, California taken by the author. Background image of the Milky Way Galaxy depicted in *The Motions of Stars Around the Galaxy* was created with the assistance of Melaina Mace (melainamace.com). Frontispiece for *Ten Million Years Aboard Spaceship Earth* is based on a NASA image from the Apollo 4 mission. Photograph of the Whirlpool Galaxy featured in *The Chilling Stars* is a Hubble mosaic published by NASA and the Milky Way seen edge-on is a mosaic of 2MASS infrared photometry published by The University of Massachusetts in collaboration with the Infrared Processing and Analysis Center (JPL/ Caltech), funded primarily by NASA and the NSF. Temple of Hephaestus appearing in *Exoplanets: Old Concept, New Science* is based on a public domain image by Sharon Mollerus. Illustration of PSR B1257+12 B & C is a remix of a NASA/JPL-Caltech illustration. Frontispiece for *Hints of Parallel Earths* is based on a NASA image from the Apollo 8 mission. The collision depicted in *When Worlds Collide* is patterned after a computer simulation by Martin Jutzi and Erik Asphaug. The nearest moon in the sky of *Water World* is based on a NASA image of Mercury from the Messenger mission. The Jovian planet appearing in the illustration *Leaving the Nest with Big Brother in Tow* is based on a NASA image of Jupiter from the Cassini mission. The segmented arthropod appearing in *Evolutionary Explosion* is based on an illustration of "Opabinia" by Nobu Tamura (palaeocritti.com) licensed under the GNU Free Documentation License. Depiction of JWST is based on a NASA illustration. Frontispiece for *The Future of Galactic Exploration* is a Hubble Space Telescope image of NGC 6384. The Soyuz-Fregat rocket depicted in *The Gaia Mission* is based on an ESA photo (Stéphane Corvaja, still iconography group).

Additional thanks to all my friends in the community of Ashland, Oregon who kept me energized throughout the course of this project and especially to Jessica Vineyard for copy-editing and proofing the entire manuscript and to Janet Sonntag and Janet Boggia for additional proofing and editorial suggestions (the author is solely responsible for typographical and stylistic errors in the final version).

Selected References

Anderson, E. and Charles, F. 2012. "XHIP: An extended Hipparcos compilation." *Astronomy Letters,* 38: 331-346.

Bahcall, J. N. and Bahcall, S. 1985. "The Sun's motion perpendicular to the Galactic plane." *Nature,* 316: 706-708.

Barbour, J. 2001. *The Discovery of Dynamics.* New York: Oxford University Press.

Barnes, R., Greenberg, R., et al. "Origin and dynamics of the mutually inclined orbits of υ Andromedae c and d." *The Astrophysical Journal,* 726: 71-77.

Berger, A. and Loutre, M. F. 2002. "An exceptionally long interglacial ahead?" *Science,* 297: 1287-1288.

de Bruijne, J. H. J. 2012. "Science performance of Gaia, ESA's space-astrometry mission." *Astrophysics and Space Science,* (in press, DOI: 10.1007/s10509-012-1019-4).

Burnham, R. 1978. *Burnham's Celestial Handbook.* New York: Dover.

Campbell, B., Walker, G. A. H., and Yang, S. 1988. "A search for substellar companions to solar-type stars." *The Astrophysical Journal,* 331: 902-921.

Fernie, J. D. 1969. "The period-luminosity relation: A historical review." *Publications of the Astronomical Society of the Pacific,* 81: 707-731.

Fernie, J. D. 1970. "The historical quest for the nature of the spiral nebulae." *Publications of the Astronomical Society of the Pacific,* 82: 1189-1230.

Francis C., Anderson E., 2012. "Evidence of a bisymmetric spiral in the Milky Way." *Monthly Notices of the Royal Astronomical Society,* 422: 1283-1293.

Francis, C. and Anderson, E. 2009. "Galactic spiral structure." *Proceedings of the Royal Society A,* 465: 3425-3446.

Gayon, J. and Bois, E. 2008. "Retrograde resonances in compact multi-planetary systems: A feasible stabilizing mechanism." *Proceedings of the International Astronomical Union (2007),* 3: 511-516.

Gillon, M., Demory, B.-O., et al. 2012. "Improved precision on the radius of the nearby super-Earth 55 Cnc e." *Astronomy & Astrophysics,* 539: A28-A34.

Greaves, J. S., Wyatt, M. C., et al. 2004. "The debris disc around τ Ceti: a massive analogue to the Kuiper Belt." *Monthly Notices of the Royal Astronomical Society,* 351: L54-L58.

Harper, G. M., Brown, A., and Guinan, E. F. 2008. "A new VLA-Hipparcos distance to Betelgeuse and its implications." *The Astronomical Journal,* 135:1430-1440.

Jutzi, M. and Asphaug, E. 2011. "Forming the lunar farside highlands by accretion of a companion moon." *Nature,* 476: 69-72.

Kant, I. 1755. *General History of Nature and Theory of the Heavens.*

Kuhn, T. S. 1962. *The Structure of Scientific Revolutions.* Chicago: Chicago University Press.

Mayor, M. and Queloz, D. 1995. "A Jupiter-mass companion to a solar-type star." *Nature,* 378: 355-359.

Moni Bidin, C., Carraro G., et al. 2012. "Kinematical and chemical vertical structure of the Galactic thick disk II. A lack of dark matter in the solar neighborhood." *The Astrophysical Journal,* 751: 30.

Parfitt, S.A., Ashton, N.M., et al. 2010. "Early Pleistocene human occupation at the edge of the boreal zone in northwest Europe." *Nature* 466, 229-233.

Pawlowski, M. S., Pflamm-Altenburg, J., and Kroupa, P. 2012. "The VPOS: A vast polar structure of satellite galaxies, globular clusters and streams around the Milky Way." *Monthly Notices of the Royal Astronomical Society,* 423: 1109-1126

Peterson, I. 1993. *Newton's Clock.* New York: Freeman.

Setiawan, J., Klement, R. J., et al. 2010. "A Giant Planet Around a Metal-Poor Star of Extragalactic Origin." *Science,* 330: 1642-1644.

Svensmark, H. and Calder, N. 2007. *The Chilling Stars.* Cambridge: Icon.

Ward, P. D. and Brownlee, D. 2000. *Rare Earth.* New York: Copernicus.

Wolszczan, A. and Frail, D. A. 1992. "A planetary system around the millisecond pulsar PSR 1257 + 12." *Nature,* 355: 145-147.

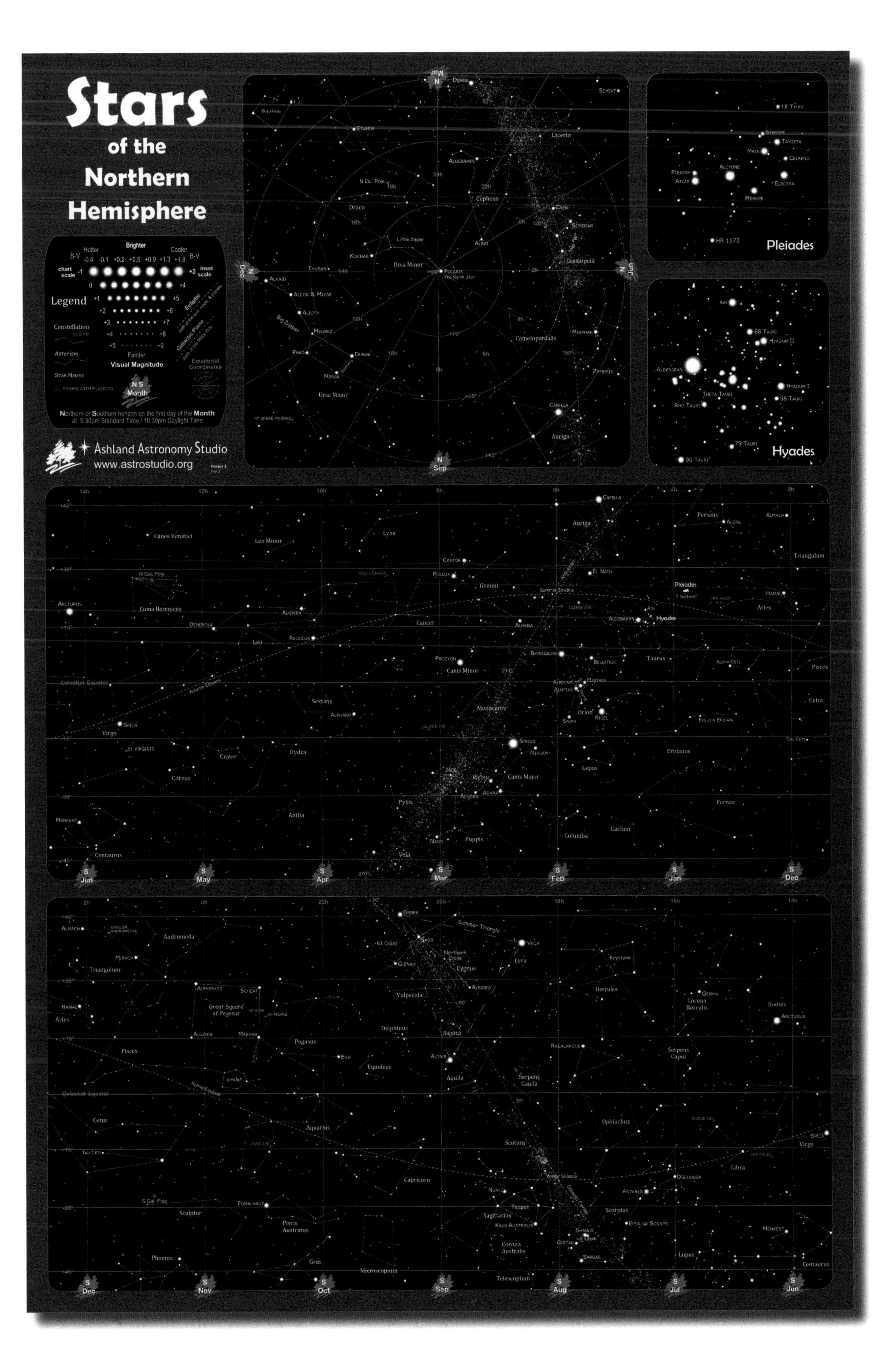

Stars
of the
Northern
Hemisphere
Legend
Brighter
Hotter
Cooler
B-V
Fainter
Visual Magnitude
chart scale
inset scale
Constellation outline
Asterism
Star Names
Ecliptic
Galactic Plane
Equatorial Coordinates
Northern or Southern horizon on the first day of the Month at 9:30pm Standard Time / 10:30pm Daylight Time
Ashland Astronomy Studio
www.astrostudio.org
Pleiades
Hyades
Ursa Minor
Little Dipper
Polaris
Draco
Cepheus
Cassiopeia
Camelopardalis
Perseus
Auriga
Capella
Ursa Major
Big Dipper
Lacerta
Orion
Taurus
Gemini
Canis Major
Canis Minor
Leo
Virgo
Hydra
Cancer
Monoceros
Eridanus
Lepus
Puppis
Columba
Vela
Pyxis
Antlia
Crater
Corvus
Centaurus
Coma Berenices
Canes Venatici
Leo Minor
Lynx
Aries
Triangulum
Pisces
Cetus
Fornax
Caelum
Sextans
Arcturus
Spica
Regulus
Procyon
Sirius
Betelgeuse
Rigel
Aldebaran
Castor
Pollux
Celestial Equator
Andromeda
Pegasus
Great Square of Pegasus
Cygnus
Northern Cross
Summer Triangle
Lyra
Vega
Deneb
Altair
Aquila
Vulpecula
Delphinus
Sagitta
Equuleus
Hercules
Corona Borealis
Boötes
Serpens Caput
Serpens Cauda
Ophiuchus
Scutum
Sagittarius
Teapot
Scorpius
Antares
Libra
Lupus
Capricorn
Aquarius
Piscis Austrinus
Fomalhaut
Sculptor
Phoenix
Grus
Microscopium
Telescopium
Corona Australis